Probability and Statistical Models

Probability and Statistical Models

Editor

Giorgos Michel

scitus
academics

Probability and Statistical Models
Edited by **Giorgos Michel**

ISBN: 978-1-68117-451-8
Library of Congress Control Number: 2016937610

SCITUS Academics LLC,
www.scitusacademics.com
Box No. 4766, 616 Corporate Way,
Suite 2, Valley Cottage,
NY 10989

Printed in United States of America on Acid Free Paper

Preface

Computational geometry is a branch of computer science dedicated to the study of geometry algorithms. Some purely geometric problems arise from the study of geometric computer algorithms, which are also part of computer geometry. Despite the recent development of modern computer geometry, it is one of the oldest computer fields in ancient history. The main impetus for the development of computer geometry as a discipline was progress in computer graphics and computer assisted design and production (CAD/CAM), but many computer geometry problems are classic in nature and can result from mathematical visualization. Other important applications for computer geometry include robotics, geographical information systems, integrated circuit design, and computer engineering and computer visualization.

This book provides a thorough introduction to algorithmic geometry and its applications. This book is intended for advanced graduates in mathematics and informatics as well as for engineering students interested in computer geometry.

Table of Contents

Chapter 1

Fluctuation-Model-Based Discrete Probability Estimation for Small Samples

Takashi Isozaki

Sony Computer Science Laboratories, Inc., Tokyo, Japan

ABSTRACT

A robust method is proposed for estimating discrete probability functions for small samples. The proposed approach introduces and minimizes a parameterized objective function that is analogous to free energy functions in statistical physics. A key feature of the method is a model of the parameter that controls the trade-off between likelihood and robustness in response to the degree of fluctuation. The method thus does not require the value of the parameter to be manually selected. It is proved that the estimator approaches the maximum likelihood estimator at the asymptotic limit. The effectiveness of the method in terms of robustness is demonstrated by experimental studies on point estimation for probability distributions with various entropies.

KEYWORDS

Probability Estimation, Fluctuation Models, Helmholtz Free Energy, Canonical Distributions

1. INTRODUCTION

For categorical observational data analysis, it is often necessary to deal with multivariate systems since variables of such data generally depend on each other. Highly predictive statistical inference requires parameters that achieve low-entropy, so it is preferable to use data of many variables because the following relationship in Shannon entropies H of random variables X and Y holds if X and Y depend on each other: $H(X|Y) \leq H(X)$ [1] , where $H(X)$ denotes Shannon entropy of X and $H(X|Y)$ denotes the conditional entropy of X given Y. These are respectively defined with marginal, joint, and conditional probability mass functions P as follows [1] :

$$H(X) = -\sum_{i} P(X=i) \log P(X=i),$$

$$H(X|Y) = -\sum_{i,j} P(X=i, Y=j) \log P(X=i|Y=j),$$

where i and j are indices of discrete states of X and Y. When we estimate probabilities in discrete probabilistic models with many-variables (e.g., Markov network and Bayesian network models [2]), statistical estimation of conditional and joint probabilities often needs exponentially large data because of the combinatorial explosion of binding events in variables. Therefore, the models are often inferred from insufficient data. The maximum likelihood (ML) method provides an estimated probability function, which of X with k discrete states is expressed by

$$P(X=i)=\frac{n_i}{n}, \quad i=1,\cdots,k$$

$$\sum_{i=1}^{k} P(X=i)=1,$$

where n and n_i respectively denote a sample size and the frequency of occurrences in i state. The estimated probability is correct in the large sample limit. However, the ML methods suffer from short size data, and few robust methods have been investigated for such data in estimation of discrete probability functions as far as we know, although many robust methods for outliers such as M-estimators have been developed [3] [4] in para- metric continuous distributions. The maximum entropy method [5] , which may be applied to small datasets, was originally appropriate for data with missing information.

In the present study, a new robust method is proposed for estimating discrete probability functions for small samples. The method uses a parameterized objective function based on Kullback-Leibler divergence [6] and Shannon entropy. The function has a similar form to the Helmholtz free energy function that appears in statistical physics [7] [8] . A key feature of the method is a model of the parameter that controls the trade-off between likelihood and robustness in response to the degree of data fluctuation. The method thus does not require the value of the parameter to be manually selected. This model is a modification of a preceding work [9] , in which the parameter is represented by an artificial model containing a free hyperparameter.

In the domain of machine learning, although several methods slightly similar to ours have been proposed [10] - [14] , there is a critical distinction between these methods and ours. Many studies that have applied free energy to statistical inference have not included the similar trade-off parameter or have treated it as a fixed value, a manually controlled parameter, or a free parameter. Regarding the existing methods, we thus consider that the potentials of free-energy-like functions have not been well extracted. Other similar methods have been developed in the context of robust estimation for outliers, in which a free parameter is introduced in an analogous fashion [3] [15] [16] . However, the problem of how to determine the value of the free parameters remains.

This paper is organized as follows. In the next section, an objective function with parameter β is introduced for robust estimation, and then probability functions obtained by the proposed method are shown to be formally equivalent to the canonical distributions that appear in statistical physics. In Section 3, a new representation of β is presented as a data-fluctuation-model, and the preferable asymptotic property of β is proved. In Section 4, some characteristic properties among quantities used in the proposed method are provided. In Section 5, we perform experiments using the proposed probability estimation method. In Section 6, conclusions regarding the estimation method are given.

2. PROBABILITY ESTIMATION WITH PARAMETER B

Note that in this paper a capital letter (such as X) denotes a random discrete variable, a non-capital letter (such as x) denotes the special state of that variable, a bold capital letter (such as $\boldsymbol{Y}$) denotes a set of variables, and a bold non-capital letter (such as $\boldsymbol{y}$) denotes configurations of that set.

To construct a method for estimating finite-discrete-probability distributions of random variable X from sample set of finite size n, the following quantities are defined. $P(X)$ denotes a discrete-probability function estimated by a proposed method that is described below. A function of X is defined as $U_0(X)$ on the basis of the Kullback--Leibler (KL) divergence [6] between empirical functions and $P(X)$ as follows:

$$U_0(X) = D\left(P(X) \| \tilde{P}(X)\right) = \sum_x P(x) \log \frac{P(x)}{\tilde{P}(x)}, \quad (1)$$

where $\tilde{P}(X)$ is a empirical distribution function. For non-parametric discrete distributions, the empirical distributions are equivalent to relative frequencies; i.e., they are equivalent to the maximum likelihood (ML) dis- tributions. $\tilde{P}(X)$ can thus be replaced by ML distributions denoted by $\hat{P}(X)$. An objective function $F(X)$ is defined as follows:

$$F(X) = U_0(X) - \frac{1}{\beta_0(X)} H(X), \quad (2)$$

where $U_0(X)$ is defined by Equation (1), $H(X)$ is the Shannon entropy [1] of the estimated functions given as $H(X) = -\sum_x P(x) \log P(x)$, and $\beta_0(X)$ is a parameter that is defined so that $0 < \beta_0(X) < \infty$. $U(X)$ and $\beta(X)$ are introduced for later convenience. $U(X)$ is represented by a cross entropy $H\left(P(X), \hat{P}(X)\right)$ and $\beta(X)$ is a normalization parameter of $\beta_0(X)$ as follow:

$$U(X) = H\left(P(X), \hat{P}(X)\right) = -\sum_x P(x) \log \hat{P}(x) \quad (3)$$

and

$$\beta(X) = \beta_0(X) / \left(1 + \beta_0(X)\right), \quad (4)$$

where $0 < \beta(X) < 1$. $F(X)$ can be rewritten by using $U(X)$ and $\beta(X)$ as:

$$F(X)=U(X)-\frac{1}{\beta(X)}H(X). \tag{5}$$

In addition, the following quantity $\epsilon(x)$ is defined as

$$\epsilon(x)=-\log\hat{P}(x). \tag{6}$$

$U(X)$ is also written with ϵ as $U(X)=\langle\epsilon\rangle(X)$; that is, $\langle\rangle$ denotes an expectation value in respect to $P(X)$.

The estimator of probability functions, $P(X)$, is defined so as to minimize Lagrangian L consisting of $F(X)$ as $\partial L/\partial P(X)=0$. L is expressed as

$$L=F+\lambda\left(\sum_x P(x)-1\right)=\frac{1}{\beta}\sum_x P(x)\log P(x)-\sum_x P(x)\log\hat{P}(x)+\lambda\left(\sum_x P(x)-1\right), \tag{7}$$

where λ is the Lagrange multiplier. $P(X)$ is thereby obtained as

$$P(x)=\frac{\exp\left(-\beta\left(-\log\hat{P}(x)\right)\right)}{\sum_{x'}\exp\left(-\beta\left(-\log\hat{P}(x')\right)\right)}=\frac{\exp\left(-\beta\epsilon(x)\right)}{\sum_{x'}\exp\left(-\beta\epsilon(x')\right)}, \tag{8}$$

where $\epsilon(x)$ as expressed in Equation (6) is used. Equation (8) is equivalent to a form known as the canonical distribution, which is also called Gibbs distribution, in statistical physics. The following equivalent form is more convenient for practical use:

$$P(x)=\frac{\left[\hat{P}(x)\right]^{\beta}}{\sum_{x'}\left[\hat{P}(x')\right]^{\beta}}. \tag{9}$$

For estimating conditional and joint-probability functions, conditional entropy $H(X|\boldsymbol{Y})$, which is defined as: $H(X|\boldsymbol{Y}) = -\sum_{x,y} P(x,\boldsymbol{y}) \log P(x|\boldsymbol{y})$, and conditional KL divergence:

$$D\left(P(X|\boldsymbol{Y}) \| Q(X|\boldsymbol{Y})\right) = \sum_{x,y} P(x,\boldsymbol{y}) \log\left(P(x|\boldsymbol{y}) / Q(x|\boldsymbol{y})\right)$$

are used. β for $P(x|\boldsymbol{y})$ is defined as

$$\beta(X|\boldsymbol{Y}) = \beta_0(X|\boldsymbol{Y}) / \left(\beta_0(X|\boldsymbol{Y}) + 1\right).$$

The formula for estimating conditional probabilities is therefore obtained by using the conditional entropy and KL divergence and $\beta(X|\boldsymbol{Y})$ in the following form:

$$P(x|\boldsymbol{y}) = \frac{\exp\left(-\beta\left(-\log \hat{P}(x|\boldsymbol{y})\right)\right)}{\sum_{x'} \exp\left(-\beta\left(-\log \hat{P}(x'|\boldsymbol{y})\right)\right)}. \quad (10)$$

Joint probability can be calculated by using Equations (8) and (10) and the definite relation $P(X,Y) = P(X|Y)P(Y)$. In general, it is calculated using decomposition rules such that

$$P(X_1, X_2, \cdots, X_n) = P(X_n|X_{n-1}, \cdots, X_2, X_1) \cdots P(X_2|X_1) P(X_1).$$

3. MODEL OF B

$P(X)$ approaches $\hat{P}(X)$ when β in Equation (9) approaches 1. On the other hand, $P(X)$ approaches the uniform distribution if $\hat{P}(X) > 0$ and β approaches 0. If β close to 1 represents that the data size is sufficiently large and if β close to 0 represents that the data size is very small, β has favorable properties for accurate and robust estimation. This is because the ML estimators generally have preferable consistency and asymptotic efficiency and the distributions close to the uniform can be regarded as the ones that have robustness for small size data. Before β is defined, the following quantity $P_n^G(X)$ with n data size is defined by a geometric mean as

$$P_n^G(X) = Z_n^G \left(\prod_{i=0}^{n} P_i(X) \right)^{\frac{1}{n+1}}, \tag{11}$$

where Z_n^G denotes a normalization constant, and $P_i(X)$ denotes the estimated function obtained from Equation (9) with initial i data. It is defined that $P_0^G(X) = P_0(X) = 1/|X|$, where $|X|$ denotes the number of states of variable X. Fluctuation δ of X with n data is given as

$$\delta_n(X) = \log P_{n-1}^G(X) - \log \hat{P}(X).$$

Then, $\beta_0(X)$ for n data is defined by using an expectation value of $\delta_n(X)$ with respect to $P_{n-1}^G(X)$, i.e., the following KL divergence:

$$\beta_0(X) = 1/D\left(P_{n-1}^G(X) \| \hat{P}(X)\right) = 1\Big/\sum_x P_{n-1}^G(x) \log \frac{P_{n-1}^G(x)}{\hat{P}(x)}, \tag{12}$$

where $n \geq 1$, and $\hat{P}(X)$ is the ML estimator function obtained from n data. It is assumed that

$D\left(P_{n-1}^{G}(X)\middle\|\hat{P}(X)\right)>0$. The normalized β_0, that is, β, is defined by Equation (4). Note that the canonical distribution $P(x)$ expressed by Equation (9) can be determined, without any free parameters, by using Equations (9), (11), (12), (4), and $P(X)=1/|X|$ for data size $n=0$. It is also defined that in Equation (10) for conditional data size $n=0$ given $Y=y$, $P(x|y)=1/|X|$ for any y in the same manner as $P(x)$.

Objective function F is rewritten in the same form as that in statistical mechanics as follows:

$$F=U-\frac{1}{\beta}H=-\frac{1}{\beta}\log Z, \tag{13}$$

where Z is the partition function, which is a similar function well known in statistical mechanics, defined for single or multivariate probabilities as

$$Z(\boldsymbol{X})=\sum_{\boldsymbol{x}}\left[\hat{P}(\boldsymbol{x})\right]^{\beta}, \tag{14}$$

and for conditional probabilities as

$$Z(X|\boldsymbol{Y})=\sum_{x}\left[\hat{P}(x|\boldsymbol{Y})\right]^{\beta}. \tag{15}$$

A significant feature of β is confirmed as follows. The lemma needed for this proof is stated as

Lemma 1.

$P_i(X)$ is denoted as the canonical distribution estimated from Equation (8) with i data. For data size $n \to \infty$, $P_n^G(x)$, defined by Equation (11), converges to a definite value $P^G(x)$ when $P_i(x) > 0$ for inte- gers i such that $i \geq 0$ and any state x of X.

Proof.

$$\begin{aligned}
&\log P_n^G(x) - \log P_{n-1}^G(x) \\
&= \frac{1}{n+1}\sum_{i=0}^{n}\log P_i(x) - \frac{1}{n}\sum_{i=0}^{n-1}\log P_i(x) + \log Z_n^G - \log Z_{n-1}^G \\
&= \left(\frac{n}{n+1} - 1\right)\log P_{n-1}^G + \frac{1}{n+1}\log P_n(x) + \log Z_n^G - \log Z_{n-1}^G.
\end{aligned}$$

(16)

Because $P_{n-1}^G(x) > 0$ and $P_n(x) > 0$, $\log P_{n-1}^G(x)$ and $\log P_n(x)$ are definite values. Thus both terms on the right-hand side of Equation (16) converge to $\log Z_n^G - \log Z_{n-1}^G$, which is a constant if $n \to \infty$. Hence, $P_n^G(x) = bP_{n-1}^G$, with constant b, can be written. Therefore, $b = 1$ and $P_n^G(x) \to P^G(x)$ in order that $P_n^G(x)$ converges not to 0 or 1 but to definite values.

Theorem 1. At the asymptotic limit (i.e., large sample limit), β converges to 1 when $P_i(x) > 0$ for integers i such that $i \geq 0$ and any state x, where $P_i(x)$ is denoted as the canonical distribution represented by Equation (8) from i data.

Proof. According to Lemma 1, $P_n^G(x) \to P^G(x)$ at the limit $n \to \infty$, where $P^G(x)$ is a definite value for any x. The following Equation is thus obtained from Equation (11) with $P_i(x)$ as follows:

$$\log P^G(x) = \lim_{n\to\infty} \frac{1}{n+1} \sum_{i=0}^{n} \log P_i(x) + \text{constant}, \tag{17}$$

$P_n(x)$ thus converges to a definite value, and $P^G(x)$ converges to $P_n(x)$ (and the constant in Equation (17) goes to 0) at $n \to \infty$. Meanwhile, ML estimator $\hat{P}(x)$ converges to true distribution $P_t(x)$ due to the consistency of the ML estimators, and β thus converges to a definite value according to Equation (9). $P_n(x)$ at $n \to \infty$ is denoted as $P_\infty(x)$. Therefore, the following Equation is derived from Equations (4), (9) and (12) at $n \to \infty$,

$$\frac{1}{\beta_0} = \frac{1-\beta}{\beta} = D\left(P_\infty(X) \| P_t(X)\right) = \sum_x \frac{\left[P_t(x)\right]^\beta}{Z} \log \frac{\left[P_t(x)\right]^{\beta-1}}{Z} = \text{a definite value} \tag{18}$$

for $P(x) > 0$ and any state x. Equation (18) requires $\beta \to 0$ or $\beta \to 1$ in order that $\left[P_t(x)\right]^\beta$ or $\left[P_t(x)\right]^{\beta-1}$ is a constant for any probability distribution $P_t(x)$. However, $\beta \to 0$ does not satisfy Equation (18), while $\beta \to 1$ satisfies it. Accordingly, $\beta \to 1$ at the asymptotic limit.

According to Theorem 1, the more data are obtained, the more β approaches 1, and the more the estimator approaches the ML estimator. The estimator that is obtained by the proposed method therefore has the same preferable asymptotic properties, namely, consistency and efficiency, as the ML estimators have. For insuffi- cient data size, β_0 is probably small due to the influence of the uniform distributions given by Equation (12), so β is also small. The estimated probability functions by the proposed method are thus interpreted as adap- tively tempered ML estimator functions in response to the degree of data fluctuation. The proposed estimation method does thereby not require manually selecting the value of parameter β, and is called "ATML," which is abbreviated as the "adaptively tempered ML" method. ATML has an advantage of simpleness over methods that need complicated algorithms (e.g., [17]).

The role of β can be seen as a trade-off parameter between likelihood and robustness by referring to another expression of Equation (2) as follows:

$$F = \beta D\left(P(X)\middle\|\hat{P}(X)\right) + (1-\beta) D\left(P(X)\middle\|P_u(X)\right) + \text{constant}, \tag{19}$$

where $P_u(X)$ denotes the uniform distribution function, which contributes to the robustness, while $\hat{P}(X)$ contributes to the likelihood. Additionally, objective function F can also be interpreted as a KL-based diver- gence measure since β is also represented by a KL divergence.

ATML has an analogy with statistical physics, since the canonical distribution for the estimator is obtained from Equation (8). Actually, U, H, β, and F respectively play similar roles to (internal) energy, entropy, (inverted) temperature, and Helmholtz free energy in statistical physics. Solving Equation $\partial L / \partial P(X) = 0$ for obtaining $P(X)$ mathematically corresponds to employing the minimum-free-energy (MFE) principle [7] in thermal physics.

ATML may seem analogous to Jaynes' maximum entropy (ME) methods [5] , which are well known as least- biased inference methods. However, the constraints on which ME methods are based may not be reliable for small samples and thus may be biased, although this kind of bias is not usually considered. On the other hand, ATML is thus designed so that even the bias can be corrected by using parameter β .

4. CHARACTERISTIC PROPERTIES OF ATML

The canonical distribution expressed as Equation (8) can provide some characteristic properties of ATML, which are similar to those in statistical physics. The following notations are defined for later convenience. Pro- bability mass functions that are estimated by ATML, denoted by P_k, have discrete states denoted as index k. The corresponding ML estimator is denoted by $\hat{P}_k$, and $\{\beta,U\}$ is used instead of $\{\beta_0,U_0\}$.

In statistical physics, (inverted) temperature β is usually defined as [7] [8]

$$\beta := \frac{\partial H}{\partial U}. \tag{20}$$

If the canonical distribution of P_k, which takes the form of Equation (8), is used, Equation (20) is auto- matically satisfied as follows:

Lemma 2. $H = \beta U + \log Z$ under the MFE condition, where H, U, β, and Z are defined in the previous sections.

Proof. Since probability mass function P_k has a canonical form under the MFE condition, it follows that

$$H = -\sum_k P_k \log P_k = -\sum_k \frac{\hat{P}_k^\beta}{Z} \log \frac{\hat{P}_k^\beta}{Z}$$
$$= \beta \cdot \left(-\sum_k \frac{\hat{P}_k^\beta}{Z} \log \hat{P}_k \right) + \left(\sum_k \frac{\hat{P}_k^\beta}{Z} \right) \cdot \log Z = \beta U + \log Z.$$

(21)

Theorem 2. Equation (20) is automatically satisfied under the MFE condition.

Proof. Partially differentiating both sides of Equation (21) with respect to U gives

$$\frac{\partial H}{\partial U} = \beta + U \frac{\partial \beta}{\partial U} + \frac{\partial}{\partial U} \log Z. \tag{22}$$

$$\frac{\partial}{\partial U} \log Z = \frac{\partial \beta}{\partial U} \frac{\partial}{\partial \beta} \log Z. \tag{23}$$

$$\frac{\partial}{\partial \beta} \log Z = \frac{1}{Z} \frac{\partial}{\partial \beta} \left(\sum_k \hat{P}_k^\beta \right) = \frac{1}{Z} \sum_k \hat{P}_k^\beta \log \hat{P}_k = \sum_k \frac{\hat{P}_k^\beta}{Z} \log \hat{P}_k = -U.$$

It follows that

$$\frac{\partial H}{\partial U} = \beta.$$

In the same way, it can be proved that $\beta_0 = \partial H / \partial U_0$.

$\langle \epsilon^2 \rangle - \langle \epsilon \rangle^2$, which is called energy fluctuations in statistical mechanics, is shown to have the following relation, where $\langle \rangle$ denotes an expectation value with respect to the canonical distributions.

$$\langle \epsilon^2 \rangle - \langle \epsilon \rangle^2 = -\frac{\partial U}{\partial \beta}. \tag{24}$$

In regard to ϵ defined in the proposed estimation method, namely, Equation (6), the same relation as that shown here is satisfied as follows:

$$\begin{aligned}
-\frac{\partial U}{\partial \beta} &= \sum_k \left(\log \hat{P}_k \right) \left\{ \frac{1}{Z} \hat{P}_k^\beta \log \hat{P}_k - \left(\frac{1}{Z^2} \right) \sum_m \frac{\partial \hat{P}_m^\beta}{\partial \beta} \right\} \\
&= \sum_k \left(\log \hat{P}_k \right) \left\{ \frac{\hat{P}_k^\beta}{Z} \log \hat{P}_k - \frac{\hat{P}_k^\beta}{Z} \sum_m \frac{\hat{P}_m^\beta}{Z} \log \hat{P}_m \right\} \\
&= \sum_k \frac{\hat{P}_k^\beta}{Z} \log \hat{P}_k \left(\log \hat{P}_k - \sum_m \frac{\hat{P}_m^\beta}{Z} \log \hat{P}_m \right) \\
&= \langle \epsilon^2 \rangle - \langle \epsilon \rangle^2,
\end{aligned}$$

Equation (24) is therefore proved.

Fisher information $\tilde{I}(\beta)$ with a parameter β is defined in the usual way as

$$\tilde{I}(\beta) := \sum_k f_k(\beta) \left(\frac{\partial}{\partial \beta} \log f_k(\beta) \right)^2, \tag{25}$$

where f is the likelihood function. We define tempered Fisher information $I(\beta)$ as Fisher information where the likelihood function is replaced with the canonical distributions with parameter β. It is shown that $I(\beta)=\langle\epsilon^2\rangle-\langle\epsilon\rangle^2$ as follows:

$$\begin{aligned} I(\beta) &= \sum_k \frac{\hat{P}_k^\beta}{Z}\left(\frac{\partial}{\partial\beta}\log\frac{\hat{P}_k^\beta}{Z}\right)^2 = \sum_k \frac{\hat{P}_k^\beta}{Z}\left(\log\hat{P}_k - \frac{1}{Z}\sum_m \hat{P}_m^\beta \log\hat{P}_m\right)^2 \\ &= \sum_k \frac{\hat{P}_k^\beta}{Z}\left(-\log\hat{P}_k\right)^2 - \left(\sum_k \frac{\hat{P}_k^\beta}{Z}\left(-\log\hat{P}_k\right)\right)^2 = \langle\epsilon^2\rangle-\langle\epsilon\rangle^2. \end{aligned} \tag{26}$$

The tempered Fisher information is therefore identical to Equation (24).

It is noteworthy that ATML has other mathematical similarities with statistical physics. That is, the same relationships that appear in statistical physics listed as follows hold.

The following relation is easily derived from the definition of partition function Z:

$$U = -\frac{\partial}{\partial\beta}\log Z. \tag{27}$$

The following relation, known as the Gibbs-Helmholtz relation, is derived from Equations (13) and (27) as

$$U = \frac{\partial}{\partial\beta}(\beta F). \tag{28}$$

The following relation is simply obtained from Equations (5) and (28) as

$$H = \beta^2 \frac{\partial F}{\partial \beta}. \tag{29}$$

The tempered Fisher information is represented by the second-order differential of the partition function for β as

$$I(\beta) = \langle \epsilon^2 \rangle - \langle \epsilon \rangle^2 = \frac{\partial^2}{\partial \beta^2} \log Z. \tag{30}$$

5. EXAMPLES

Numerical experiments are performed to demonstrate the robustness of ATML for small samples, in comparison with the ML and ME methods [5] .

X is assumed to have three internal states and four probability mass functions with a variety of entropies denoted as $H(X)$in natural logarithms as

1. $P(x=0)=0.431, P(x=1)=0.337, P(x=2)=0.232, H(X)=1.07$,
2. $P(x=0)=0.677, P(x=1)=0.206, P(x=2)=0.117, H(X)=0.841$,
3. $P(x=0)=0.851, P(x=1)=0.117, P(x=2)=0.0320, H(X)=0.498$,
4. $P(x=0)=0.9898, P(x=1)=0.00810, P(x=2)=0.00210, H(X)=0.0621$.

Data from each function was sampled, and probabilities were estimated from given data sets with various data sizes. According to convention, averaged outputs $\langle X \rangle$ were set as the constraint in the ME method as follows:

$\langle X \rangle := (1/N)\sum_{d=1}^{N} X_d$, where Xd denotes d-th sample's output, and N denotes sample size. After that, true and estimated probabilities were compared by using KL divergence as a metric with the following form:

$$D\left(P(X) \| P_e(X)\right) = \sum_x P(x) \log \frac{P(x)}{P_e(x)}, \tag{31}$$

where $P(X)$ is the true distribution, and $P_e(X)$ is the distribution estimated by ML, ME, or ATML. For avoiding zero probabilities, probabilities were smoothed by adding 0.0001 to the counts.

The KL divergences are shown in Figure 1, where they are averaged values from 100 samples at each sample size from identical distributions. It can be seen that the ML estimators are inferior to ATML due to overfitting, except for the distribution having very small entropy. Even the degree of superiority of the ML estimation in (d) is relatively smaller than that of inferiority in other distributions. The ME estimators showed the opposite be- haviors to those by the ML estimators, and showed some relatively poor results in large-sample regions. ML methods tend to fit data and can thus more accurately estimate distributions with very low entropies than others in small-sample cases. For example, if a true entropy equals zero, the ML method can estimate the exact true distribution from only one sample. On the other hand, ME methods tend to increase entropies and can thereby accurately estimate distributions with high entropies close to the uniform distributions. Hence, the ML method tends to overfit data, and the ME method tends to underfit data in the view of misestimation. Even so, ATML showed relative stability in terms of both sample sizes and distributions. This result indicates the effectiveness of ATML as a probability estimation method.

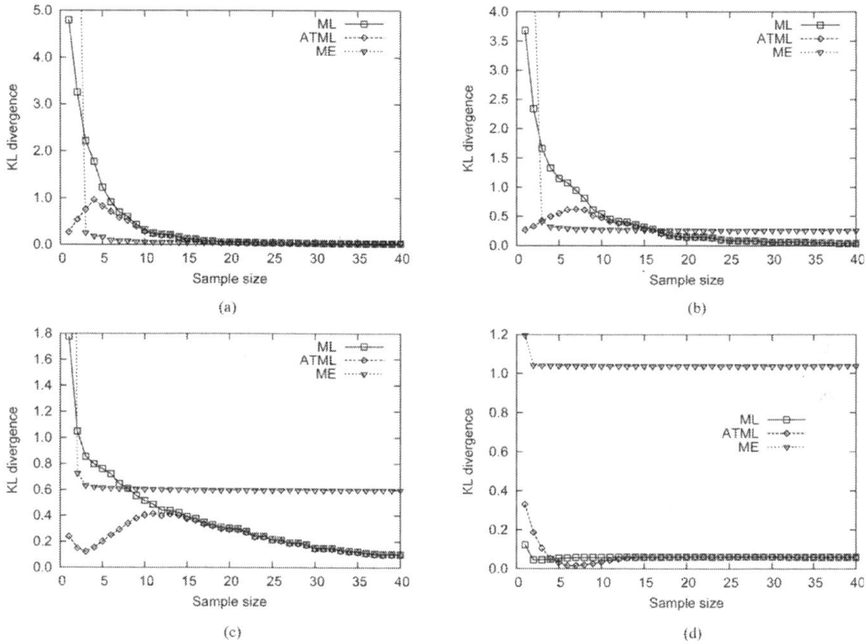

Figure 1. KL divergences between true probability mass functions and probability mass functions estimated by using ML, ATML, and ME. The horizontal axes denote sample sizes. H denotes Shannon entropy in natural logarithms. (a) H = 1:07; (b) H = 0:841; (c) H = 0:498; (d) H = 0:0621.

6. CONCLUSION

A robust method for estimating discrete probability functions, called "adaptively tempered maximum likelihood" method (ATML for short), is proposed. The estimators obtained in this method minimize a parameterized objective function similar to Helmholtz free energies that appear in statistical physics. The key feature of the proposed method is a model of the parameter as a fluctuation of finite size data. The parameter that is modeled plays an important role in determining the appropriate trade-off between likelihood and robustness in response to the degree of the fluctuations. ATML does thereby not require manually selecting the value of the parameter. It is also proved that the obtained estimator approaches the maximum likelihood estimator at the asymptotic limit. The effectiveness of ATML in terms of robustness was demonstrated by experimental studies on point estimation for probability distributions with various entropies.

ACKNOWLEDGEMENTS

The author thanks Mario Tokoro and Hiroaki Kitano of Sony Computer Science Laboratories, Inc. for their support as well as Te Sun Han, Hiroshi Nagaoka, Tomohiro Ogawa and Jun Suzuki of the University of Electro- Communications for their valuable comments and discussions.

REFERENCES

1. Shannon, C.E. (1948) A Mathematical Theory of Communication. Bell Systems Technical Journal, 27, 379-423, 623-656.

2. Pearl, J. (1988) Probabilistic Reasoning in Intelligent Systems. Morgan Kaufmann, San Mateo, CA.

3. Basu, A., Harris, I.R., Hjort, N.L. and Jones, M.C. (1998) Robust and Efficient Estimation by Minimising a Density Power Divergence. Biometrika, 85, 549-559. http://dx.doi.org/10.1093/biomet/ 85.3. 549

4. Beran, R. (1977) Minimum Hellinger Distance Estimates for Parametric Models. Annals of Statistics, 5, 445-463.

5. Jaynes, E.T. (1957) Information Theory and Statistical Mechanics. Physical Review, 106, 620-630.

6. Kullback, S. and Leibler, R.A. (1951) On Information and Sufficiency. Annals of Mathematical Statistics, 22, 79-86.

7. Callen, H.B. (1985) Thermodynamics and an Introduction to Thermostatistics. 2nd Edition, John Wiley & Sons, Hoboken, NJ.

8. Kittel, C. and Kroemer, H. (1980) Thermal Physics. W. H. Freeman, San Francisco, CA.

9. Isozaki, T., Kato, N. and Ueno, M. (2009) "Data Temperature" in Minimum Free Energies for Parameter Learning of Bayesian Networks. International Journal on Artificial Intelligence Tools, 18, 653-671.

10. Hofmann, T. (1999) Probabilistic Latent Semantic Analysis. Proceedings of Conference on Uncertainty in Artificial Intelligence (UAI-99), Stockholm, 30 July-1 August 1999, 289-296.

11. LeCun, Y. and Huang, F.J. (2005) Loss Functions for Discriminative Training of Energy-Based Models. Proceedings of International Workshop on Artificial Intelligence and Statistics (AISTATS-05), Barbados, 6-8 January 2005, 206-213.

12. Pereira, F., Tishby, N. and Lee, L. (1993) Distributional Clustering of English Words. In: Proceedings of Annual Meeting on Association for Computational Linguistics (ACL-93), Association for Computational Linguistics, Stroudsburg, 183-190.

13. Ueda, N. and Nakano, R. (1995) Deterministic Annealing Variant of the EM Algorithm. Proceedings of Advances in Neural Information Processing Systems 7 (NIPS 7), Denver, 29 November-1 December 1994, 545-552.

14. Watanabe, K., Shiga, M. and Watanabe, S. (2009) Upper Bound for Variational Free Energy of Bayesian Networks. Machine Learning, 75, 199-215.

15. Jones, M.C., Hjort, N.L., Harris, I.R. and Basu, A. (2001) A Comparison of Related Density-Based Minimum Divergence Estimators. Biometrika, 88, 865-873.

16. Windham, M.P. (1995) Robustifying Model Fitting. Journal of the Royal Statistical Society B, 57, 599-609.

17. Pöschel, T., Ebeling, W., Frömmel, C. and Ramírez, R. (2003) Correction Algorithm for Finite Sample Statistics. The European Physical Journal E, 12, 531-541.

Chapter 2

Dirichlet Compound Multinomials Statistical Models

Paola Cerchiello, Paolo Giudici

Department of Economics and Management, University of Pavia, Pavia, Italy

ABSTRACT

This contribution deals with a generative approach for the analysis of textual data. Instead of creating heuristic rules for the representation of documents and word counts, we employ a distribution able to model words along texts considering different topics. In this regard, following Minka proposal (2003), we implement a Dirichlet Compound Multinomial (DCM) distribution, then we propose an extension called sbDCM that takes explicitly into account the different latent topics that compound the document. We follow two alternative approaches: on one hand the topics can be unknown, thus to be estimated on the basis of the data, on the other hand topics are determined in advance on the basis of a predefined ontological schema. The two possible approaches are assessed on the basis of real data.

KEYWORDS

Textual Data Analysis; Mixture Models; Ontology Schema; Reputational Risk.

1. INTRODUCTION

With the rapid growth of on-line information, text categorization has become one of the key techniques for handling and organizing data in textual format. Text categorization techniques are an essential part of text mining and are used to classify new documents and to find interesting information contalned within sevcral on-line web sites. Since building text classifiers by hand is difficult, time-consuming and often not efficient, it is worthy to learn classifiers from experimental data. In this proposal we employ a generative approach for the analysis of textual data. In the last two decades many interesting and powerful contributions have been proposed. In particular, when coping with the text classification task, a researcher has to face the well-known problem of polysems (multiple senses for a given words) and synonyms (same meaning for different words). One of the first effective model able to solve those issues is represented by Latent semantic analysis (LSA) [1]. The basic idea is to work at a semantical level by reducing the vector space through Singular Value Decomposition (SVD), producing not sparse occurrence tables that help in discovering associations between documents. In order to establish a solid theoretical statistical framework in this context, in [2] a probabilistic version of LSA (pLSA) has been proposed, also known as the aspect model, rooted in the family of latent class models and based on a mixture of conditionally independent multinomial distributions for the couple words-documents. The intention from the introduction of pLSA was to offer a formal statistical framework, helping the parameter interpretation issue as well. By the way the goal was achieved only partially, in fact the multinomial mixtures, which components can be interpreted as topics, offer a probabilistic justification at words but not at documents level. In fact the latter are represented merely as list of mixing proportions derived from mixture components. Moreover, the multinomial distribution presents as many values as there are in the training documents and therefore it learns topic mixture on those trained documents. The extension to previously unseen documents is not appropriate since there can be new topics. In order to overcome the asymmetry between words and documents and to produce a real generative model, [3] proposed the LDA (Latent Dirichlet Allocation). The idea of such new approach emerges from the concept of exchangeability for the words in a document that unfolds in the "bag of words" assumption: the order of words in a text is not important. In fact the LDA model is able to capture either the words or documents exchangeability unlike LSA and pLSA. On the other hand LDA is a generative model in any sense since it posits a Dirichlet distribution over documents in the corpus, while each topic is drawn from a Multinomial distribution over words. However note that [4] in 2003 have shown that LDA and pLSA are equivalent if the latter is under a uniform Dirichlet prior distribution. Obviously LDA does not solve all the issues. The main restriction embedded in LDA approach and due to the Dirichlet distribution, is the assumption of independence among topics. The immediate consequence was to tackle the issue by introducing the Correlated Topic Model (CTM), as proposed in [5]. CTM introduces correlations among topics by replacing the Dirichlet random variable with the logistic normal distribution. Unlike LDA, CTM presents a clear complication in terms of inference and parameter estimation since the logistic normal distribution and the Multinomial are not conjugate. To bypass the problem, the most recent alternative is represented by the Independent Factor Topic Models (IFTM) introduced in [6]. Such proposal makes use of latent variable model approach to detect hidden correlations among topics. The choice to explore the latent model world allows to choose among several alternatives ranging from the type of relation, linear or not linear, to the type of prior to be specified for the latent source. For sake of completeness is important to mention another interesting research path focusing on the burstiness phenomenon, that is the tendency of rare words, mostly, to appear in burst. The above mentioned generative models are not able to capture such peculiarity, that instead is very well modelled by the Dirichlet Compund Multinomial model (DCM). Such distribution was introduced by statisticians [7] and has been widely employed by other sectors like bioinformatics [8] and language engineering [9]. An important contribution in the context of text classification was brought by [10] and [11] that profitably used DCM as a bag-of-bags-of-words generative process. Similarly to LDA, we have a Dirichlet random variable that generates a Multinomial random variable for each document from which words are drawn. By the way, DCM cannot be considered a topic model in a way, since each document derives specifically by one topic. That is the main reason why [12] proposed a natural extension of the classical topic model LDA by plugging into it the DCM distribution and obtaining the so called DCMLDA. Following this line of thinking, we move from DCM approach and we propose an extension of the DCM, called "semanticbased Dirichlet Compound Multinomial" (sbDCM), that permits to take latent topics into account. The paper is organized as follows: in Section 2 we show the Dirichlet Compound Multinomial (DCM) model, in Section we propose an extension of the DCM, called "semanticbased Dirichlet Compound Multinomial" (sbDCM), in Section 4 we show how to estimate the parameters of the different models. Then, in Section 5 we assess the predictive performance of the two distributions by using seven different classifiers. Finally we show the different classification performance according to the knowledge on the topics T (known or unknown).

2. BACKGROUND: THE DIRICHLET COMPOUND MULTINOMIAL

The DCM distribution is a hierarchical model: on one hand, the Dirichlet random variable is devoted to model the Multinomial word parameters q; on the other hand, the Multinomial variable models the word count vectors $\overline{x}$ comprising the document. The distribution function of the DCM mixture model is:

$$p(\overline{x}|\alpha)=\int_{\theta}p(\overline{x}|\theta)p(\theta|\alpha)\mathrm{d}\theta \quad (1)$$

where $p(\overline{x}|\theta)$ is the Multinomial distribution:

$$p(\overline{x}|\theta)=\frac{n!}{\prod_{w=1}^{W}x_w}\prod_{w=1}^{W}\theta_w^{x_w}. \quad (2)$$

in which $\overline{x}$ is the words count vector, x_w is the count for each word and q_w the probability of emitting a word w_i; therefore a document is modelled as a single set of words ("bag-of-words"). The Dirichlet distribution $p(\theta|\alpha)$ is instead parameterized as:

$$p(\theta|\alpha)=\frac{\Gamma\left(\sum_{w=1}^{W}\alpha_w\right)}{\prod_{w=1}^{W}\Gamma(\alpha_w)}\prod_{w=1}^{W}\theta_w^{\alpha_w-1}. \quad (3)$$

with $\alpha=\{\alpha_w\}$ the Dirichlet parameter vector for words, as consequence the whole set of words ("bag-of-bags") is modelled. Thus a text (a document in a set) is modelled as “bag-of-bags-of-words”. Developing the previous integral we obtain:

$$p(\theta|\alpha)=\frac{n!}{\prod_{w=1}^{W}(x_w)}\frac{\Gamma\left(\sum_{w=1}^{W}\alpha_w\right)}{\Gamma\left(\sum_{w=1}^{W}(x_w+\alpha_w)\right)}\prod_{w=1}^{W}\frac{\Gamma(x_w+\alpha_w)}{\Gamma(\alpha_w)} \tag{4}$$

In Figure 1 we report the graphical representation of the DCM model. From another point of view, each Multinomial is linked to specific sub-topics and makes, for a specific document, the emission of some words more likely than others. Instead the Dirichlet represents a general topic that compounds the set of documents and thus the DCM could be also described as “bag-of-scaleddocuments”. The added value of the DCM approach consists in the ability to handle the “burstiness” of a rare word without introducing heuristics [13]. In fact, if a rare word appears once along a text, it is much more likely to appear again.

When we consider the entire set of documents$^{(D)}$, where each document is independent and identified by its count vector, $\left(D=\{\overline{x}_1,\overline{x},\cdots,\overline{x}_N\}\right)$, the likelihood of the whole documents set (D) is

$$p(D|\alpha)=\prod_{d=1}^{N}p(\overline{x}_d|\alpha)$$

$$=\prod_{d=1}^{N}\left(\frac{\Gamma\left(\sum_{w=1}^{W}\alpha_w\right)}{\Gamma\left(x_d+\sum_{w=1}^{W}\alpha_w\right)}\prod_{w=1}^{W}\frac{\Gamma(x_w+\alpha_w)}{\Gamma(\alpha_w)}\right) \tag{5}$$

where x_d is the sum of the counts of each word in the document d-th (åx_{dw}) and x_{dw} the count of word w-th for the document d-th. Thus the log-likelihood is:

$$\log\left(p\left(D|\alpha\right)\right)$$
$$=\sum_{d=1}^{N}\log\Gamma\left(\sum_{w=1}^{W}\alpha_w\right)-\log\Gamma\left(x_w+\sum_{w=1}^{W}\alpha_w\right)$$
$$+\sum_{d=1}^{N}\sum_{w=1}^{W}\log\Gamma\left(x_{dw}+\alpha_w\right)-\log\Gamma\left(\alpha_w\right) \tag{6}$$

The parameters can be estimated by a fixed-point iteration scheme, as described in Section 4.

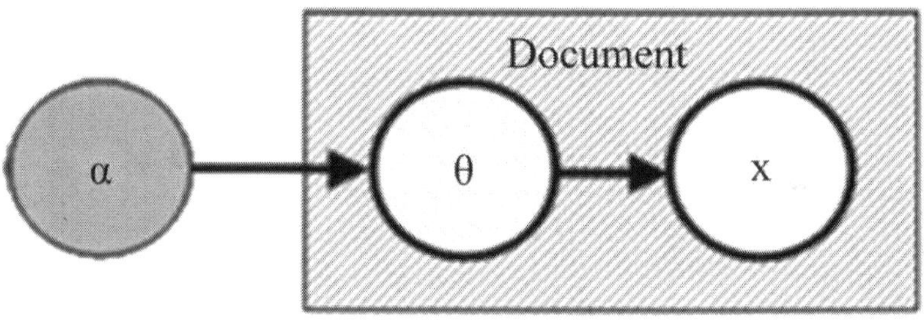

Figure 1. Hierarchical representation of DCM model.

3. A SEMANTIC-BASED DCM

As explained in Section 2, we have a coefficient a_w for each word compounding the vocabulary of the set of documents which is called "corpus". The DCM model can be seen as a "bag-of-scaled-documents" where the Dirichlet takes into account a general topic and the Multinomial some specific sub-topics. Our aim in this contribution is to build a framework that allows us to insert specifically the topics (known or unknown) that compound the document, without losing the "burstiness" phenomenon and the classification performance. Thus we introduce a method to link the a coefficients to the hypothetic topics, indicated with $\beta=\{\beta_i\}$, by means of a function $\alpha=F(\beta)$ which must be positive in b since the Dirichlet coefficients are positive. Note that usually $\dim(\mathrm{b}) \ll \dim(\mathrm{a})$ and, therefore, our proposed approach is parsimonious. Substituting the new function into the integral in Equation (1), the new model is:

$$p(\overline{x}|\beta) = \int_{\theta} p(\overline{x}|\theta) p(\theta| F(\beta)) \mathrm{d}\theta, \quad (7)$$

We have considered as function F(b) a linear combination based on a matrix D and the vector $\overline{\beta}$. D contains information about the way of splitting among topics the observed count vectors of the words contained in a diagonal matrix A and $\overline{\beta}$ is a vector of coefficient (weights) for the topics. More specifically we assume that:

$$A = \begin{pmatrix} w_1 & & \\ & \ddots & \\ & & w_W \end{pmatrix}, D = \begin{pmatrix} d_{11} & \cdots & d_{1T} \\ \vdots & \ddots & \vdots \\ d_{W1} & \cdots & d_{WT} \end{pmatrix},$$

$$\beta = \begin{pmatrix} \beta_1 \\ \vdots \\ \beta_t \\ \vdots \\ \beta_T \end{pmatrix}, D^* = A \times D \quad (8)$$

$$F(\overline{\beta}) = D^* \times \overline{\beta} = \alpha = \begin{pmatrix} \alpha_1 \\ \vdots \\ \alpha_t \\ \vdots \\ \alpha_T \end{pmatrix} \quad (9)$$

Note that:

- $\alpha_w = \sum_t^T d_t^* \beta_t$ with T the number of Topics;
- $d_{wt}^* = w_w \times d_{wt}$;
- d_{wt} is the coefficient for word w-th used to define the degree of belonging to topic t-th and by which a portion of the count of word w-th is assigned to that particular topic t-th.

By substituting this linear combination into Equation (4), we obtain the same distribution but with the above mentioned linear combination for each a:

$$p(\bar{x}|\beta) = \frac{n!}{\prod_{w=1}^{W}(x_w)} \frac{\Gamma\left(\sum_{w=1}^{W}\sum_t^T d_{wt}^* \beta_t\right)}{\Gamma\left(\sum_{w=1}^{W}\left(x_w + \sum_t^T d_{wt}^* \beta_t\right)\right)} \prod_{w=1}^{W} \frac{\Gamma\left(x_w + \sum_t^T d_{wt}^* \beta_t\right)}{\Gamma\left(\sum_t^T d_{wt}^* \beta_t\right)} \tag{10}$$

This model is a modified version of the DCM, henceforth semantic-based DCM (sbDCM), and the new loglikelihood for the set of documents becomes:

$$\log(p(D|\beta)) = \sum_{d=1}^{N} \log\Gamma\left(\sum_{w=1}^{W}\sum_t^T d_{wt}^* \beta_t\right) - \log\Gamma\left(x_w + \sum_{w=1}^{W}\sum_t^T d_{wt}^* \beta_t\right) + \sum_{d=1}^{N}\sum_{w=1}^{W}\left(\log\Gamma\left(x_{dw} + \sum_t^T d_{wt}^* \beta_t\right) - \log\Gamma\left(\sum_t^T d_{wt}^* \beta_t\right)\right) \tag{11}$$

In Figure 2 we report the graphical representation of the new model where the a's are substituted by a function of the b's. An important aspect of the proposed approach is represented by the number T of topics to be inserted into the sbDCM that can be:

- Unknown, thus to be estimated on the basis of the available data.
- A priori known (i.e. fixed by field experts).

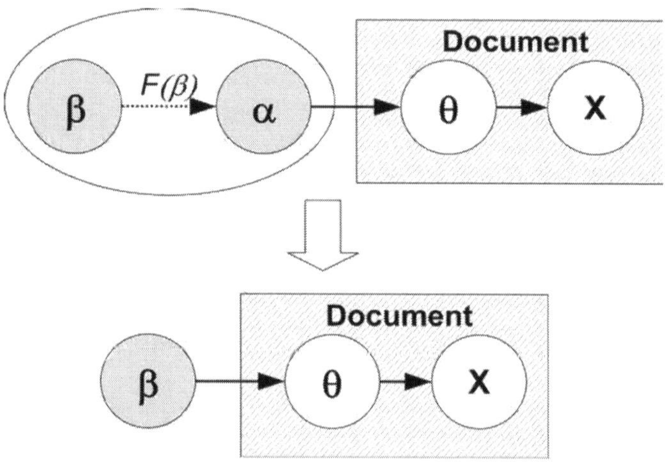

Figure 2. Hierarchical representation of sbDCM model.

The first case will be treated in Section 5.1, in particular since it is not always possible to know in advance the number of latent topics present in a corpora, it becomes very useful to build a statistical methodology for discovering efficiently and consistently a suitable number of topics. In this case the number of topics T can be considered a random variable. Thus, we use a segmentation procedure to group the words in order to create groups of words sharing common characteristics that can be considered as latent topics. The analysis is completed by choosing the best number of groups and a distance matrix is used to set the membership percentage (d_{wt}) of each word to each latent topic. The second case will be treated in Section 5.2 and proposes to exploit the sbDCM model by employing a priori knowledge based on ontological schemas that describe the relations among concepts with regards to the general topics of the corpora. The ontology structure provides the set of relations among the concepts to which can be associated a certain number of key words, by a field expert(s). Thus, we want to use the classes of a given ontology and the associated key words to define in advance the number T of topics.

4. PARAMETERS ESTIMATION

There are several methods to maximize the log-like- lihood and to find the parameters associated with a DCM: simplified Newton iteration, the Expected Maximization method and the maximization of the simplified likelihood (called "leave-one-out" likelihood, LOO). Among them, the most general and flexible algorithm is the Expected Maximization (EM). The EM algorithm is an iterative procedure able to compute the maximum-likelihood estimates whenever data are incomplete or they are considered complete but not observable (latent variables) [14].

In general, considering $p(X,Z|\theta)$the joint probability distribution (or probability mass function) where X is the incomplete (but observable) data set and Z the missing (or latent) data, we can calculate the complete maximum likelihood estimation of model parameters q by means of the EM algorithm. We obtain an overall complexity reduction for what concerns the calculation of observed maximum-likelihood (ML) estimates. The goal is achieved by alternating the expectation (E-step) of the likelihood with the latent variables as if they are observed and the maximization (M-step) of the likelihood function found in the E-step. With the result of the M-step we update the new parameters q_{new} that are used again in the cycle, by starting from the E-step, until we reach a fixed degree of approximation e of the observed likelihood which increases step by step moving towards the maximum (that could be local) [15]. The EM can be built in different ways. One possibility is to see the EM as a lower bound maximization where we alternate the E-step to calculate an approximation of the lower bound for the log-likelihood and maximize it in the M-step until a stationary point (zero gradient) is reached. If we are able to find a lower bound for the log-likelihood we can maximize it via a fixed-point iteration; in fact it is the same principle of considering the EM as a lower bound maximization. In our context, for the DCM, the lower bound with $\log(p(D|\alpha))$ is the following quantity:

$$\alpha_w^{k+1} = \alpha_w^k \frac{\sum_w^k \left[\psi\left(x_{dw} + \alpha_w^k\right) - \psi\left(\alpha_w^k\right)\right]}{\sum_d^N \left[\psi\left(x_d + \sum_w^W \alpha_w^k\right) - \psi\left(\sum_w^W \alpha_w^k\right)\right]} \tag{12}$$

this allows us to use a fixed point iteration whose steps are:

$$\alpha_w^{k+1} = \alpha_w^k \frac{\sum_w^k \left[\psi\left(x_{dw} + \alpha_w^k\right) - \psi\left(\alpha_w^k\right)\right]}{\sum_d^N \left[\psi\left(x_d + \sum_w^W \alpha_w^k\right) - \psi\left(\sum_w^W \alpha_w^k\right)\right]} \tag{13}$$

where x_d is the sum of the counts of each word in the document d-th $\left(\sum x_{dw}\right)$, x_{dw} the count of word w-th for the document d-th and α_w^k the Dirichlet coefficient for word w at the k-th step. The algorithm is stopped when a degree of approximation e is reached. The iteration starts with a_w equals to the occurrence percentage of the word w-th in the corpus. The estimated parameters, as said before, have an important characteristic: they follow the "burstiness" phenomenon of words. In fact the smaller a_w is, the more "burstiness" effect is contained within a word, as revealed in Section 5.

In the case of sbDCM by considering the new loglikelihood, Equation (11), we can use the same lower bound employed before. The only modification lays on the substitution of the a coefficients with the linear function of b, as in Equation (9), and thereby the new fixed point iteration step is:

$$\beta_t^{k+1} = \beta_t^k \frac{\sum_d^N \left[\sum_w^W \left(d_{wt}^* \psi \left(x_{dw} + \sum_t^T d_{wt}^* \beta_t^k \right) \right) - \sum_w^W \left(d_{wt}^* \right) \psi \left(\sum_t^T d_{wt}^* \beta_t^k \right) \right]}{\sum_d^N \left[\sum_w^W \left(d_{wt}^* \right) \psi \left(x_{dw} + \sum_w^W \sum_t^T d_{wt}^* \beta_t^k \right) - \sum_w^W \left(d_{wt}^* \right) \psi \left(\sum_w^W \sum_t^T d_{wt}^* \beta_t^k \right) \right]} \tag{14}$$

As before we stop the iteration at a fixed degree of approximation and the coefficients d_{wt} are those described in Section 3. The new a's maintain the words burstiness, as we shall show in Section 5 and they are used to classify the document by employing a Naive Bayes classifier. For our applications, in the next section we have used for both models a value of e of 10^{-10}. The iteration starts with b_t equals to the percentage of each single cluster obtained by the grouping analysis of vocabulary words.

5. MODEL PERFORMANCE

In this section we describe the evaluation of the different classifiers by using the parameters estimated from the DCM distribution. Thus, our training data set is compound of 6436 documents with a vocabulary (already filtered and stemmed) of 15,655 words so we have to estimate 15,655 a parameters. The a parameter is able to model the "burstiness" of a word. In fact, the smaller the a parameters are, the more bursty the emission of words is. This phenomenon is characteristic of rare words, therefore a coefficients are, on average, smaller for less counted words. The average value of the overall a parameters is 0.0342, the standard deviation is 0.1087 and the maximum and minimum values are respectively 6.6074 and 0.0025. Once the coefficient vector of a's is obtained we employ seven different classifiers, three of which are described in [13] (normal (N), complement (C) and mixed (M). The remaining ones are proposed as the appropriate combination of the previous ones, in order to improve their characteristics. Those new classifiers are set in function of the number of words that a test-document has in common with the set of documents that compound a class; in this way we create a classifier in function of the number of words in common. Thus we analyze the following additional classifiers: COMPLEMENT + MIXED + NORMAL (CMN), COMPLEMENT + NORMAL (CN), COMPLEMENT + MIXED (CM), MIXED + NORMAL (MN).

In order to evaluate the classification performance we employ three performance indexes:

• Ind1: The proportion of true positive over the total number of test-documents:

$$\left(\sum_{d=1}^{D}\frac{TP_d}{D}\right)\times 100;$$

• Ind2: The proportion of classes that we are able to classify:

$$\left(\sum_{c=1}^{C}\frac{I_c}{C}\right)\times 100;$$

where I_c is an indicator that we set 1 if at least one document of the class is classified correctly, otherwise we set 0.

• Ind3: The proportion of true positive within each class over the number of test documents present in the class:

$$\left(\frac{1}{C}\sum_{c=1}^{C}\frac{TP_c}{M_c}\right)\times 100;$$

where M_c is the number of test-documents in each class, TP_c is the number of true positive in the class and C the number of topics (46).

For the four combined classifiers such indexes have been calculated by varying the number of words in common between the test document and the class. In particular for our test we have used three different thresholds for the number of words (n): 15, 10 and 5. For example, we indicate with the initials CM_n the classification rule that employs classifier C when the number of common words are less or equal to n and classifier M when the number of words in common is more than n. Instead, the initial $CMN_{n.m}$ identifies the using of classifier C until n, the classifier N over m and the classifier M between n and m. For the data at hand the number of words in common between the two sets (training and evaluation set) varies between 1 and 268. The above mentioned combination is based on the following idea: if the number of words in common between the bag of words and the correct class is low, then the most information content is in the complement set. Otherwise the needed information is contained either in the normal set or in the complement one. Taking into account such consideration we have set up different combination and we concluded that the useful trade-off among classifiers is equal to 10 (Table 1).

As we can see the best classifiers are the mixed and the CM_{10} ones. They are able to classify respectively 1237 and 1238 over 1609 documents that are distributed not uniformly among classes (46). These classifiers are able to classify at least a document per class even if there are classes containing only 2 documents. Between them the CM_{10} classifier has index three slightly better than mixed one. The worse classifier, in this case, is the complement version alone.

We now verify the goodness of the sbDCM models described in Section 3, to understand if we can insert latent topics into DCM by maintaining the burstiness and the same classification performance. Two kinds of matrixes have been used in the cluster procedure. One matrix contains the correlations among words C in the vocabulary and another G is constructed by calculating the Kruskal-Wallis index on the count matrix among words. The latter index g is defined as follows:

+(15)

where n_i is the number of sample data, N the total observation number of the k samples, k the number of samples to be compared and $\overline{r_i}$ the mean rank of i-th group. The denominator of the index g is a correction factor needed when tied data are present in the data set, where p is the number of recurring ranks and c_i is the times the i-th rank is repeated. The index g depends on the differences among the averages of the groups $\overline{r}_i$ and the general average. If the samples come from the same population or from populations with the same central tendency, the arithmetic averages of the ranks of each group $\overline{r}_i = \sum_j r_{ij} / n_i$ should be similar to each other and to the general average $(N+1)/2$ as well. The training dataset contains 2051 documents with a vocabulary of 4096 words for both approaches. The evaluation dataset (again the same for both models) contains 1686 documents which are distributed over 46 classes. In Tables 1 and 2 we report the results from the two tests obtained respectively by matrixes C and G and by varying the number of groups in the cluster. In the tables we indicate with LL_{in} the log-likelihood before the parameters updating and with LL_{out} after the iteration procedure which is stopped when the error e = 10^{-10} is reached. The same with AIC_{cin} and AIC_{cout} that is the corrected Akaike Information Criterion (AIC_c) before and after the uploading. The indexes Ind1, Ind2 and Ind3 have been described in Section 5.1.

As we can see in the two Tables 1 and 2, the percentages of correct classification (Ind1) are very close to the original ones with a parameter for each word (4096 parameters). Of course they depend on the type of classifier employed during the classification step. Considering both sbDCM and DCM, the differences produced by varying the number of groups are small. Moreover the AIC_c is always better in the new approach then considering each word as a parameter (DCM model). In particular for what concerns the approach based on the correlation matrix C (in Table 2) with 17 groups and on the Mixed classifier, it can predict correctly the 68.43% of documents. The log-likelihood and the AIC_c indexes along groups are quite similar, however the best value is obtained with 5 groups (respectively −205,412 and 410,834). Considering again the approach based on the correlation matrix C, we can conclude that, in terms of complexity expressed by the AIC index, the sbDCM approach whatever applied classifier is always better than the DCM. When we use matrix G (Table 2) the best classification performance is for the complement classifier based on 23 groups, with a percentage of 68.72%, a log-likelihood of −204,604, the AIC_c of 409,254. The best log-likelihood and AIC_c are for cluster with 46 groups (respectively −204,362 and 408,816). Even if the sbDCM distribution based on matrix G is not able to improve the classification performance of DCM, we can say that the sbDCM Index1 is always very close to the best one. In addition the new model is always better in terms of either AIC and log-likelihood indexes. Moreover, if we perform an asymptotic chisquared test (χ^2_{test}) considering the two cases (matrixes G and C) to decide whether the difference among loglikelihoods (LL), with respect to DCM, are significant (i.e. the difference is statistically meaningful if the $|LL_1 - LL_2|$ is greater than 6), we can see from Tables 1 and 2 the test with matrix G has the best performance.

Performance of the Semantic-Based Dirichlet Compound Multinomial with T Known in Advance A different approach needs to be assessed when the number of available topic T is known in advance. In fact a text corpora could be enriched by several descriptions of treated topics according to the knowledge of field experts. In more details, the analysis could be provided with a priori knowledge based on ontological schemas that describe the relations among concepts with regards to the general topics of the corpora. An ontology (from which an ontological schema is derived) is a formal representation of a set of concepts within a domain and the relationships between those concepts [16]. It provides a shared vocabulary, which can be used to model a domain, that is, the type of objects and/or concepts that exist, and their properties and relations. In Figure 3 we report an example of graphical representation of an ontological schema. For example, if a text set deals with reputational risk management for corporate institutions, an ontology can be created on the basis of the four categories of problems (internal processes, people, systems and external events) defined by Basel II Accords.

Table 1. Classification results by varying cluster numbers and using matrix C.

Classifier	Measure	*sbDCM5*	*sbDCM11*	*sbDCM17*	*sbDCM23*	*sbDCM46*	*DCM*
	LL_i	-282,226	-265,250	-252,197	-247,125	-242,991	-222,385
	LL_o	-205,412	-205,614	-205,601	-205,597	-205,602	-205,286
	AIC_i	56,446	530,522	504,228	494,296	486,074	454,264
	AIC_o	410,834	411,250	411,236	411,240	411,296	420,066
Norm.	*Ind1*	68.13	68.13	67.95	68.19	68.25	68.78
=	*Ind2*	97.83	97.83	97.83	97.83	97.83	97.83
=	*Ind3*	62.32	62.32	62.15	62.32	62.32	61.61
Comp	*Ind1*	68.19	68.31	68.25	68.37	68.25	68.78
=	*Ind2*	100	100.0	100.0	100.0	100.0	100.0
=	*Ind3*	66.20	66.30	66.05	66.58	66.01	67.89
Mixed	*Ind1*	68.07	68.36	68.43	68.37	68.31	68.07
=	*Ind2*	97.83	97.83	97.83	97.83	97.83	97.83
=	*Ind3*	63.87	64.01	64.05	64.05	64.01	63.87

Table 2. Classification results by varying cluster numbers and using matrix G.

Classifier	Measure	*sbDCM5*	*sbDCM11*	*sbDCM17*	*sbDCM23*	*sbDCM46*	*DCM*
	LL_i	−291,257	−283,294	−270,360	−266,453	−258,061	−222,385
	LL_o	−205,912	−204,647	−204,600	−204,604	−204,362	−205,286
	AIC_i	582,524	566,610	540,754	532,952	516,214	454,264
	AIC_o	411,834	409,316	409,234	409,254	408,816	420,066
Norm.	*Ind1*	67.83	67.71	67.47	67.42	67.65	67.66
〃	*Ind2*	97.83	97.83	97.83	97.83	97.83	97.83
〃	*Ind3*	62.02	61.73	61.45	61.43	61.55	61.61
Comp	*Ind1*	67.95	68.66	68.55	68.72	68.60	68.78
〃	*Ind2*	100	100.0	100.0	100.0	100.0	100.0
〃	*Ind3*	67.95	68.09	68.05	68.29	68.05	67.89
Mixed	*Ind1*	68.07	68.13	67.83	67.71	67.89	68.07
〃	*Ind2*	97.83	97.83	97.83	97.83	97.83	97.83
〃	*Ind3*	63.87	63.97	63.05	62.86	62.95	63.87

Hence we can suppose that some specific sub-topics and key words, such as the possible causes of reputetional losses, will be almost surely treated along the texts.

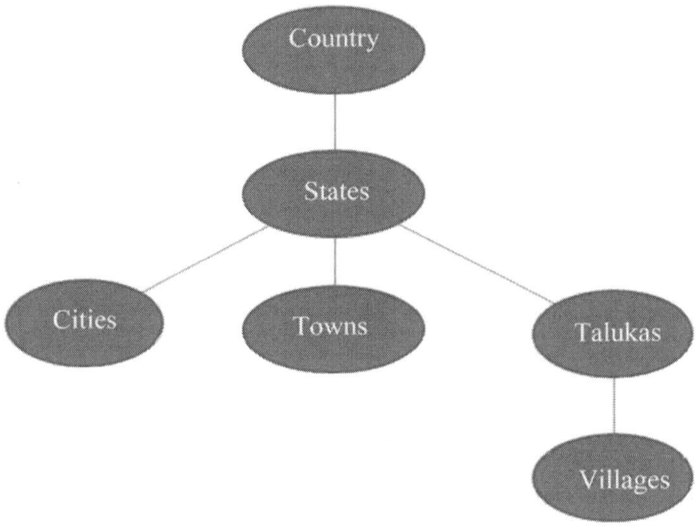

Figure 3. Example of ontological schema.

Thereby, the ontology structure provides the set of relations among the concepts to which can be associated, by a field expert(s), a certain number of key words. Thus, we want to use the classes of a given ontology and the associated key words to define in advance the number T of topics. The ontological schema which we refer to, for the data at hand, deals with the so called reputational risk [17]. It is not simple to define and consequently to measure and to monitor the reputation concept since it involves intangible assets such as: honor, public opinion, perception, reliability, merit. By the way, it is a matter of fact, that a bad reputation can seriously affect and condition the performance of a company. Moreover companies tend to act once the adverse event has occurred. According to such approach we can say that there is not a risk management activity, but only a crisis management. With regards to reputational risk, media coverage plays a key role in determining a company's reputation. This often occurs when a company reputation has been significantly damaged by unfair attacks from special interest groups or inaccurate reporting by the media. A detailed and structured analysis of what the media are saying is especially important because the media shape the perceptions and expectations of all the involved actors. Natural language processing technologies enable these services to scan a wide range of outlets, including newspapers, magazines, TV, radio, and blogs. In order to enable the application of the classification textual model sbDCM we have collaborated with the Italian market leader company in financial and economic communication, Sole24ORE. Sole- 24ORE team provided us with a set of 870 articles about Alitalia, an Italian flight company, covering a period of one year (Sept 07-Sept 08).

The 80% of the articles are used to train the model and the remaining 20% to validate the process. The objective is to classify the articles on the basis of the reputation ontology in order to understand the argument treated in the articles. The ontology classes used for the classification are:

- Identity: the perception that stakeholders have of the organization, person, product. It describes how the organization is perceived by the stakeholders.
- Corporate Image: the "persona" of the organitation. Usually for companies visibly manifested by way of branding and the use of trademarks and involves the mission and the vision. It involves brand value.
- Integrity: personal inner sense of "wholeness" deriving from honesty and consistent uprightness of character.
- Quality: the achievement or excellence of an entity. Quality is sometimes certificated by a third part.

- Reliability: ability of a system to perform/maintain its functions in routine and also in different hostile or/ and unexpected circumstances. It involves customer satisfaction and customer fidelitation.

- Social Responsibility: social responsibility is a doctrine that claims that an organization or individual has a responsibility to society. It involves foundation campaign and sustainability

- Technical Innovation: the Introduction of new technical products or services. Measure of the "RD orientation" of an organization (only for companies).

- Value For Money: the extent to which the utility of a product or service justifies its price.

Those classes define the concept of reputation of a company. To link the ontology classes to the textual analysis we use a set of key words for each class of the reputation schema. Since the articles are in Italian, the key words are in Italian as well. For example the concept of "Reliability" involves customer satisfaction and customer fidelitation and is characterized by the following set of key words: affidabilità, fiducia, consumatori, risorsa, organizzazione, commerciale, dinamicità, valore, mercato. On the basis of these key words, we perform a grouping analysis considering the 9 classes. From the clustering, we derive the matrix D. Empirical results show good performance in terms of correct classification rate. We obtain that, given 171 articles, we correctly classify the 68% of them.

6. CONCLUSIONS

This contribution has shown how to enrich the DCM model with a semantic extension. We also have proposed a method to insert latent topics within the "Dirichlet Compound Multinomial" (DCM) without losing the words "burstiness": we call such a distribution "semanticbased Dirichlet Compound Multinomial" (sbDCM). The approaches assessed depend on the knowledge about the topics T. In fact there can be two alternative contexts: on one hand the topics are unknown in advanced, thus to be estimated on the basis of data at hand. On the other hand a text corpora could be enriched by several descriptions of treated topics according to the experience of the field expert(s). Specifically, the analysis can be empowered with a priori knowledge based on ontological schemas that describe the relations among concepts with regards to the general class argument of the corpora. In order to insert topics we create a new coefficient vector b_t for each topic and later on we obtain the a parameters as a linear combination of them. The methodology is based on a matrix D containing the degree of membership of each word to a cluster (i.e. a topic) by using the cluster distance matrix. Then we split the words count vectors among latent topics and, by employing a fixed-point iteration, we generate the β coefficients representing the topics weights. In order to compare the two models DCM and sbDCM we have employed a "Naive Bayes Classifier" based on the estimated distributions as shown in [13]. Several classifiers have been proposed and tested, and among them the best performance is obtained by means of the "mixed formula" and "CM_{10}". Moreover, we run several tests to verify if the classification performance reached with an a for each word (DCM) is maintained or improved by the sbDCM.

Such an objective has been accomplished employing two different approaches. We propose two different methods to generate b parameters, one based on the correlation among words C and the second based on the Kruskal-Wallis index calculated on the words count matrix G. The results report that the test performances in terms of misclassification rate are quite close to each other and to the performance reached by the DCM. However the sbDCM distribution is able to obtain better results in terms of AIC and log-likelihood especially in the case of matrix G. Concluding, by using matrix D to describe how words count vectors can be split among topics, the b as weights for the topic and a as a linear combination we are able to obtain an optimal classification performance and to follow the burstiness.

ACKNOWLEDGEMENTS

This work has been supported by MUSING 2006 contract number 027097, 2006-2010 and FIRB, 2006-2009). The paper is the result of the close collaboration between the authors, however the paper has been written by Paola Cerchiello under the supervision of Paolo Giudici.

REFERENCES

1. S. Deerwester, S. Dumais, G. W. Furnas, T. K. Landauer and R. Harshman, "Indexing by Latent Semantic Analysis," Journal of the American Society for Information Science, Vol. 41, No. 6, 1990, pp. 391-407.

2. T. Hofmann, "Probabilistic Latent Semantic Indexing," Proceedings of Special Interest Group on Information Retrieval, New York, 1999, pp. 50-57.

3. D. M. Blei, A. Y. Ng and M. I. Jordan, "Latent Dirichlet Allocation," Journal of Machine Learning Research, Vol. 3, 2003, pp. 993-1022.

4. M. Girolami and A. Kaban, "On an Equivalence between PLSI and LDA," Proceedings of Special Interest Group on Information Retrieval, New York, 2003, pp. 433-434.

5. D. M. Blei and J. D. Lafferty, "Correlated Topic Models," Advances in Neural Information Processing Systems, Vol. 18, 2006, pp. 1-47.

6. D. Putthividhya, H. T. Attias and S. S. Nagarajan, "Independent Factor Topic Models," Proceeding of International Conference on Machine Learning, New York, 2009, pp. 833-840.

7. J. E. Mosimann, "On the Compound Multinomial Distribution, the Multivariate B-Distribution, and Correlations among Proportions," Biometrika, Vol. 49, No. 1-2, 1962, pp. 65-82.

8. K. Sjolander, K. Karplus, M. Brown, R. Hughey, A. Krogh, I. S. Mian and D. Haussler, "Dirichlet Mixtures: A Method for Improving Detection of Weak but Significant Protein Sequence Homology," Computer Applications in the Biosciences, Vol. 12, No. 4, 1996, pp. 327- 345.

9. D. J. C. Mackay and L. Peto, "A Hierarchical Dirichlet Language Model," Natural Language Engineering, Vol. 1, No. 3, 1994, pp. 1-19.

10. T. Minka, "Estimating a Dirichlet distribution," Unpublished Paper, 2003. http://research.microsoft.com/en-us/um/people/minka/papers/dirichlet/

11. R. E. Madsen, D. Kauchak and C. Elkan, "Modeling Word Burstiness Using the Dirichlet Distribution," Proceeding of the 22nd International Conference on Machine Learning, New York, 2005, pp. 545-552.

12. G. Doyle and C. Elkan, "Accounting for Burstiness in Topic Models," Proceeding of International Conference on Machine Learning, New York, 2009, pp. 281-288.

13. J. D. M. Rennie, L. Shih, J. Teevan and D. R. Karge, "Tackling the Poor Assumptions of Naive Bayes Text Classifier," Proceeding of the 20th International Conference on Machine Learning, Washington DC, 2003, 6 p.

14. A.P. Dempster, M. N. Laird and D. B. Rubin, "Maximum Likelihood from Incomplete Data via the EM Algorithm," Journal of the Royal Statistical Society, Series B, Vol. 39, No. 1, 1977, pp. 1-38.

15. D. Böhning, "The EM Algorithm with Gradient Function Update for Discrete Mixture with Know (Fixed) Number of Components," Statistics and Computing, Vol. 13, No. 3, 2003, pp. 257-265.

16. S. Staab and R. Studer, "Handbook on Ontologies, International Handbooks on Information Systems," 2nd Edition, Springer, Berlin, 2009.

17. P. Cerchiello, "Statistical Models to Measure Corporate Reputation," Journal of Applied Quantitative Methods, Vol. 6, No. 4, 2011, pp. 58-71.

Chapter 3

Statistical Foundation of Empirical Isotherms

F. Brouers

Department of Chemical Engineering, Liege University, Liege, Belgium

ABSTRACT

We show that most of the empirical or semi-empirical isotherms proposed to extend the Langmuir formula to sorption (adsorption, chimisorption and biosorption) on heterogeneous surfaces in the gaseous and liquid phase belong to the family and subfamily of the BurrXII cumulative distribution functions. As a consequence they obey relatively simple differential equations which describe birth and death phenomena resulting from mesoscopic and microscopic physicochemical proc- esses. Using the probability theory, it is thus possible to give a physical meaning to their empirical coefficients, to calculate well defined quantities and to compare the results obtained from different isotherms. Another interesting consequence of this finding is that it is possible to relate the shape of the isotherm to the distribution of sorption energies which we have calculated for each isotherm. In particular, we show that the energy distribution corresponding to the Brouers- Sotolongo (BS) isotherm [1] is the Gumbel extreme value distribution. We propose a generalized GBS isotherm, calculate its relevant statistical properties and recover all the previous results by giving well defined values to its coefficients. Finally we show that the Langmuir, the Hill-Sips, the BS and GBS isotherms satisfy the maximum Bolzmann-Shannon entropy principle and therefore should be favoured.

KEYWORDS

Adsorption Isotherms, Burr Functions, Adsorption Energy Distribution, Maximum Entropy

1. INTRODUCTION

Every year hundreds or more papers are devoted to the analysis of sorption (physical adsorption, chemi- and bio-sorption) of gas or solutions on a variety of substrates [2] . Among them, a great number are concerned with the decontamination of air, water and soil. One of the typical procedures is a comparison of the data with empirical isotherm formulas which in the course of time have been proposed by scientists working in the field to generalize the original Langmuir isotherm to heterogeneous surfaces and to sorption in solutions. Most of these formulas are empirical and bring little information on the physicochemical processes responsible for the particular shape of the isotherm curves. The evolution of the empirical parameters with external factors is recorded but there are no precise correlations between the variations of the parameters belonging to different isotherms. It appears that some order should be introduced in that field in order to propose a more rigorous classification of the sorbent-sorbate couples.

In this paper which is a contribution to that effort, we want to emphasize that since some of these isotherms appear to be genuine cumulative probability distributions, they should be favoured, formulated in the language of the theory of probability and might bring more quantitative and more structured information making advantage of their mathematical properties. The probability theory of complex systems has made considerable progress these last years and one can expect that its introduction in the field of sorption could be of great help.

2. SORPTION ON HETEROGENEOUS SURFACES

A few years ago we published a paper [1] actualizing the efforts initiated by Langmuir, Zeldowitsch and followers eighty years ago to incorporate in the classical Langmuir adsorption isotherm theory, the heterogeneous nature of the substrate, the N-body interactions and the nonequilibrium state of the sorbate. One important conclusion of this study was that the most important ingredient playing a role in designing the shape of the isotherm is the sorption energy distribution which itself is a reflection of the disordered and complex nature of the phenomenon. In our work, we insisted on the fact that it would be useful to rewrite the theory in the framework of the theory of probability. Moreover we reminded that it is an asymmetric birth and death (sorption-desorption) process and a rare event dominated problem due to the very nature of the sorption mechanism, the more active sites being the first to be occupied. We pointed out that these characteristics should be taken into account in the theory. We showed that to account for the power law Freundlich isotherm, one has to assume a Lévy heavy tail behavior for the temperature dependent Langmuir parameter.

As a consequence of this study we proposed an isotherm using a Weibull distribution known since as Brouers-Sotolongo (BS) isotherm which has been used among others in sorption on porous/nonporous surface interface [3] , magnetic nano-particles [4] , adsorption on doped nanostructures [5] , on activated carbon produced from natural products [6] - [8] , algae [9] - [11] , soils and natural wastes [12] for water treatment [9] [13] , biosorp- tion and biodegradation, food contamination [14] as well as medical applications such as the chemical immobili- zation of bacteriophages on surfaces [15] [16] .

The present paper is a extension of some of the ideas developed in our previous works. We will take advantage of the recent progress in the statistical theory of complex and deterministic chaotic systems. We will show that many of the isotherms used in the literature, especially in the treatment of water, form a subfamily of the BurrXII distribution. This will lead us to propose a generalization (GBS) of the BS isotherm replacing the exponential in the Weibull function by a deformed exponential used now in the formulation of the nonextensive thermodynamics [17] and other complex systems theories. The same technique has helped us to elucidate the universality of relaxation in disordered systems [18] [19] and formulate a fractional-time kinetics for n-order reaction systems [20] . As we will show, many of the isotherms used in the literature can be obtained by giving well defined values to the parameters of this generalized isotherm.

3. THE BURRXII DISTRIBUTION FUNCTION

If we view the isotherm as a cumulative distribution function we can write the isotherms in the following forms:

$$\tilde{\Theta}(P(C)) = \frac{\Theta(P(C))}{\Theta_{\max}} = \int_0^{P(C)} \theta(p(c))\mathrm{d}p(\mathrm{d}c). \tag{1}$$

In Equation (1), $\theta(p(c))$ is the relative sorbed quantity as the pressure or concentration are increased in the gas or liquid phase in appropriate units. The quantity $\Theta_{\max}$ is the maximum sorption capacity in appropriate units. The $p(c)$ are supposed to be related thermodynamically to a sorption energy variable e:

$$p(c) \propto \exp(-e/RT). \tag{2}$$

In an heterogeneous system, as we increase the pressure or the concentration, the most active sites with the highest sorption energy are first occupied until complete saturation. With a change of variable, one can write

$$\tilde{\Theta}(P) = \int_{\Delta(P)} \theta(e)\,\mathrm{de} \tag{3}$$

where $\Delta(P)$ is the range of energies involved at pressure P and $\theta(e)$ is an energy dependent properly normalized distribution function. This second formulation (Equation (3)) has been used to determine an empiri- cal formula for the sorption energy distribution [21] -[24] . In the following the variables P or c will be de- noted by the greek letter $\varkappa$.

We will now demonstrate that if we choose for $\tilde{\Theta}(\varkappa)$, the $Burr_{XII}$ cumulative distribution function (cdf) many of the physically sound isotherms used in the literature to generalize the Freundlich formula can be recovered and a new generalized isotherm can be proposed as a synthesis of the efforts of a few generations.

In probability theory and statistical sciences, the $Burr_{XII}$ distribution is a continuous probability distribution for a non-negative random variable [25] . It is also known in econometrics as the Singh-Maddala distribution [26] where it has been used as a generalization of the Pareto distribution for the graduation over the whole range of incomes and is used to measure the level of inequality.

The $Burr_{XII}$ distribution is a a member of a system of continuous cumulative distribution (cdf) functions introduced by I. Burr in 1942 [25] . It has the form:

$$F_B(x,a,b,c) = 1 - \left[1 + c\left(\frac{x}{b}\right)^a\right]^{-1/c} \tag{4}$$

where a,b,c are positive parameters. Its normalized probability density function (pdf) $f_B(x,a,b,c)$ is ob- tained from

$$\mathrm{d}F_B(x,a,b,c) = f_B(x,a,b,c)\mathrm{d}x = \frac{a}{b}\left(\frac{x}{b}\right)^{a-1}\left[1 + c\left(\frac{x}{b}\right)^a\right]^{-1-1/c}\mathrm{d}x \tag{5}$$

$$\mathrm{d}F_B(x^a,a,b,c) = f_B(x^a,a,b,c)\mathrm{d}x^a = \left(\frac{1}{b}\right)^a\left[1 + c\left(\frac{x}{b}\right)^a\right]^{-1-1/c}\mathrm{d}x^a. \tag{6}$$

In previous papers [18] [19] , we have shown how it could be derived from the maximum entropy principle using a generalization of the non-extensive Tsallis entropy with appropriate constraints. However more recently, it has been shown that it can be derived more naturally from the classical Boltzmann-Gibbs entropy with appro- priate generalization of the moments constraints (see Section 9).

The cumulative distribution functions belonging to the Burr family are solution of the general differential equation

$$\frac{\mathrm{d}F(x)}{\mathrm{d}(x)} = f(x) = g(x)F(x)(1 - F(x)) \tag{7}$$

where $F(x)$ and $g(x)$ are continuous functions defined in specific domains. This differential equation describes a birth and death function modulated by a $g(x)$ function which applied to a particular problem depends on the nature of the phenomena and the influence of the environment. The first and most studied of these differential equations is the famous Verhulst logistic equation introduced in 1845 [27] to mimic and cal- culate population dynamics. In that case, $g(x)=1$ and its solution is

$$F_V(x)=1/(1+\exp(-x)). \tag{8}$$

In its discrete form it has been one of the first model of deterministic chaos [28] .

For the $Burr_{XII}$ cdf, the twelfth one in the family, the function $g(x)=\tilde{g}_B(x)/x$ has the form of an hyper- bolic type function, the function, $\tilde{g}_B(x)$ varying smoothly between a and a/c.

$$g(x)=\tilde{g}_B(x)/x \text{ where } \tilde{g}_B(x)=\frac{\left(a(x/b)^a\right)}{\left(1+c(x/b)^a\right)\left(1-\left(1+c(x/b)^a\right)^{(-1/c)}\right)}$$
$$\tilde{g}_B(x)\to a \text{ when } x\to 0 \text{ and } \tilde{g}_B(x)\to a/c \text{ when } x\to\infty \tag{9}$$

The $Burr_{XII}$ distribution function has become a reference distribution in complex and non equilibrium systems as the exponential and Gaussian distributions are the reference distributions in equilibrium and non interacting systems. The "dialectic" form of its differential equation shows that it could be useful to deal with phenomena like for instance epidemic propagation, population evolution, kinetics of complex reactions, eco- nomic evolution, pharmacokinetic, cancer remission and obviously sorption-desorption. It has been used exten- sively these last years in a variety of chaos, nonlinear and nonequilibrium problems in quasi all fields of pure and applied sciences including natural phenomena, meteorology, hydrology, earthquake, economy, sociology and medicine.

An other interesting feature of the $Burr_{XII}$ distribution is the existence of two power laws tails, one for $x \to 0$ with exponent a and one for $x \to \infty$ with exponent $\mu = a/c$. It has a limited number of finite moments depending on the value of μ. When $0 < \mu < 1$, it has a heavy tail and belongs to the basin of attraction of the family of stable Lévy distributions. It is to say, it has some peculiar properties which have interesting consequences. Lévy functions do not obey the traditional central limit theorem and an expectation value of x cannot be defined. For higher values of μ the average value increases with the number of observations following a well defined power law [29] .

4. THE SUBFAMILY OF THE BURRXII DISTRIBUTION AND THE ASSOCIATED ISOTHERMS

The $Burr_{XII}$ function (Equation (4)) can generate a sub-family of cdf distributions if one gives particular values to the two parameters a and c in $F_B(x,a,b,c)$. The case $a=1$ and $c=0$ is simply the exponential function. Some of these functions coincide with the form of well known empirical isotherms:

For $c=0$

$$F_B(x,a,b,0) = F_W(x,a,b) = 1 - \mathrm{Exp}\left(-\left(\frac{x}{b}\right)^a\right) \tag{10}$$

$$\tilde{g}_W(x \to 0) \to a \text{ and } \tilde{g}_W(x \to \infty) \to \infty. \tag{11}$$

This is a Weibull distribution. The corresponding isotherm in the sorption literature is known as the Brouers- Sotolongo (BS) isotherm:

$$\tilde{\Theta}_{BS}(\varkappa) = \left[1 - \text{Exp}\left(-\left(\frac{\varkappa}{b}\right)^a\right)\right]. \tag{12}$$

If $\varkappa \ll b$, one gets the Freundlich isotherm

$$\Theta_F(\varkappa) = K_F \varkappa^a. \tag{13}$$

If moreover one puts $a = 1$ in Equation (12), one gets the Jovanovic isotherm [30]

$$\tilde{\Theta}_J(\varkappa) = \left[1 - \text{Exp}\left(-\left(\frac{\varkappa}{b}\right)\right)\right]. \tag{14}$$

For $c = 1$, one has:

$$F_B(x,a,b,1) = F_{HS}(x,a,b) = \frac{\left(\frac{x}{b}\right)^a}{1+\left(\frac{x}{b}\right)^a} = 1 - \left[1+\left(\frac{x}{b}\right)^a\right]^{-1} \tag{15}$$

$$f_B(x,a,b,1) = f_{HS}(x,a,b) = \frac{a}{x}\frac{\left(\frac{x}{b}\right)^a}{\left(1+\left(\frac{x}{b}\right)^a\right)^2}; \quad g_{HS}(x,a,b) = ax^{-1} \tag{16}$$

which is called in probability theory the loglogistic function. The corresponding isotherms are the Hill, the Langmuir-Freundlich and Sips isotherms

$$\tilde{\Theta}_{LF}(\varkappa)=\frac{\left(\frac{\varkappa}{b}\right)^{a}}{1+\left(\frac{\varkappa}{b}\right)^{a}}=1-\left[1+\left(\frac{\varkappa}{b}\right)^{a}\right]^{-1}. \tag{17}$$

If both a and c are equal to 1:

$$F_B(x,1,b,1)=F_{HS}(x,1,b)=F_L(x,b)=\frac{x}{b+x}=1-\left[1+\left(\frac{x}{b}\right)\right]^{-1} \tag{18}$$

the corresponding isotherm is the Langmuir isotherm.

$$\tilde{\Theta}_L(\varkappa)=\frac{\varkappa}{b+\varkappa}. \tag{19}$$

As discussed in [1], the exponent a is related to the width and shape of the sorption energy distribution which itself depends on the heterogeneity of the substrate. In Section 8 we will show that it defines an effective temperature $T^{*}=T/a$.

In the isotherms we have just reviewed, the exponent a is supposed to be constant and do not change with the evolution of the sorbed quantity Θ. This is a restrictive assumption. An isotherm derived from the full $Burr_{XII}$ would allow the characteristic exponent to vary slowly from a to a/c. Therefore quite naturally a more realistic isotherm based on the full $Burr_{XII}$ distribution can be proposed:

$$\tilde{\Theta}_{GBS}(\varkappa)=1-\left[1+c\left(\frac{\varkappa}{b}\right)^{a}\right]^{-1/c}. \tag{20}$$

This generalized BS isotherm has a unified character since it contains the Langmuir, the Freundlich-Langmuir, the Hill and the Sips isotherm and as we will see in the next section, the Generalized Freundlich-Langmuir and the Toth isotherms. The GBS isotherm can be written in a more compact form

$$\tilde{\theta}_{GBS}(\varkappa)=\left\{1-\left[1+c\left(\frac{\varkappa}{b}\right)^{a}\right]^{-1/c}\right\}=1-\exp_c\left(-\left(\frac{\varkappa}{b}\right)^{a}\right). \tag{21}$$

We have used the definition of the deformed exponential function introduced in mathematics in the XIX century and appearing to day in the theory of many complex systems

$$\exp_c(x)=(1-cx)^{-\frac{1}{c}} \text{ if } 1-cx>0, \quad 0 \text{ otherwise} \tag{22}$$

$$\log_c(x)=\frac{1-x^{-c}}{c} \text{ with } \exp_c(\log_c(x))=\log_c(\exp_c(x))=x. \tag{23}$$

When $c=0$, one recovers the usual exponential. In the nonequilibriun thermodynamic literature $c=q-1$ where q is the nonextensive (nonadditive) entropy index [17] . In the complex reaction literature, $c=n-1$, where n is the effective fractional reaction order. In the extreme value theory $c=\xi$, the shape parameter of the distribution. We recover the BS isotherm θ_{BS} if $c=0$ and the Hiil-Sips adsorption for $c=1$.

This new isotherm has four parameters $\theta_{\max}, a, b,$ and c which have simple physical interpretation: $\theta_{\max}$ is the maximum saturation sorbed quantity, a is the Freundlich exponent which is related to the width and shape of the sorption energy and is a measure of the distribution. When $a<1$, it can be related to the self- similar (fractal) properties at the micro- and meso-scopic scale. For $a>1$, it has been interpreted as the mani- festation of a multi-molecular site sorption [31] . The coefficient $0 \leq c \leq 1$ is related to the cluster organization of the system. A large c corresponds to a strong clustering organization [32] . The coefficient b is a T depen- dent scale parameter and combined with a and c allows the calculation of all the quantities characterizing the statistical distribution: expectation, variance and moments, median, quantiles and some other coefficients which measure quantitatively the way the sorption depends on the concentration or the pressure. These useful expressions for the analysis of isotherms are derived in the appendix. The value $a=1$ separates the distribu- tions defining the isotherms in two groups. For $a \leq 1$ and this includes the Langmuir isotherm, the pdf is L-shaped while for $a>1$, it is unimodular. This has a strong influence on the nature of the sorption. We will show also in the appendix that the quantity b^a is directly related to specific moment of the probability distri- bution. Finally when $(a/c)=\mu$ is <1 it is the heavy tail (Lévy) exponent which controls the saturation be- havior of the sorption curve.

5. THE GENERALIZED FREUNDLICH-LANGMUIR AND TOTH ISOTHERMS

A two exponents isotherm (GFL) generalizing the Freundlich-Langmuir (Hill, Sips) isotherm was proposed by Marczewski and Jaroniec [33] .

$$\tilde{\theta}_{GFL}(\varkappa)=\left(\frac{\left(\frac{\varkappa}{b}\right)^{n}}{1+\left(\frac{\varkappa}{b}\right)^{n}}\right)^{\frac{m}{n}}. \tag{24}$$

The corresponding cdf function

$$F_G(x)=\left(\frac{\left(\frac{x}{b}\right)^{n}}{1+\left(\frac{x}{b}\right)^{n}}\right)^{\frac{m}{n}}=\left(1+\left(\frac{x}{b}\right)^{-n}\right)^{-\frac{m}{n}} \tag{25}$$

has the characteristics of a cdf $F_G(0)=0,\ F_G(\infty)=1$.

It appears that $F_G(x)$ has the form of a Dagun function [34] used concurrently with the $Burr_{XII}$ equation in econometrics. It can be related to the $Burr_{XII}$ function by a simple change of variables. This will allow us to relate the isotherms obtained from the GFL isotherm form (20) to the ones already derived.

As $F_B(x)$, $F_G(x)$ is also the solution of a first order differential equation. Indeed one has

$$\frac{\mathrm{d}F_G(x)}{\mathrm{d}x}=\frac{mb\left(\frac{\left(\frac{x}{b}\right)^{n}}{1+\left(\frac{x}{b}\right)^{n}}\right)^{\frac{m}{n}}}{x\left(1+\left(\frac{x}{b}\right)^{n}\right)}=\frac{mb}{x}\left(\frac{x}{b}\right)^{-n}F_G(x)^{\frac{n}{m}+1}. \tag{26}$$

We have moreover:

$$\text{for } x \to 0, \frac{\mathrm{d}F_G(x)}{\mathrm{d}x} \to mb\left(\frac{x}{b}\right)^{m-1} \text{ and for } x \to \infty, \frac{\mathrm{d}F_G(x)}{\mathrm{d}x} \to mb\left(\frac{x}{b}\right)^{-n-1}. \tag{27}$$

These asymptotic behaviors which are supposed to be the same as the ones of $\frac{\mathrm{d}F_B(x)}{\mathrm{d}x}$ gives the relations between the exponents of the two formulations.

$$m = a, \; m/n = c, \; n = a/c = \mu \tag{28}$$

Starting from the GLF isotherm equation, one can recover some of the empirical isotherms: for $m = n = 1$, the Langmuir isotherm, for $m = n$, the Langmuir-Freundlich or Hill isotherm. For $n = 1$, the Sips isotherm and for $m = 1$ the Toth [35] isotherm. The first ones belong to the subfamily of the $Burr_{XII}$ subfamily isotherms and have been already considered. The Toth isotherm is applicable only for $m = a = 1$ and for $c = 1/n$. We will see now how the Marczewski and Jaroniec GLF is linked to the $Burr_{XII}$ function using the relations between the two probability functions.

6. DAGUM DISTRIBUTION VERSUS BURRXII DISTRIBUTION

The BurrXII cdf and pdf functions (Equations (4), (5)) can be written

$$F_B\left(x, a, \tilde{b}, c\right) = 1 - \left[1 + \left(\frac{x}{\tilde{b}}\right)^a\right]^{-1/c} \quad \text{with } \tilde{b} = b\left(c^{-\frac{1}{a}}\right) \tag{29}$$

$$f_B\left(x,a,\tilde{b},c\right)=\frac{a}{\tilde{b}c}\left(\frac{x}{\tilde{b}}\right)^{a-1}\left[1+\left(\frac{x}{\tilde{b}}\right)^{a}\right]^{-1-1/c}.$$

(30)

If we make the change of variables $x \to 1/x$ and $\tilde{b} \to 1/\tilde{b}$ in (Equations (29) and (30)), we get the Dagum cdf and pdf:

$$F_D\left(x,a,\tilde{b},c\right)=\left[1+\left(\frac{x}{\tilde{b}}\right)^{-a}\right]^{-1/c}$$

(31)

$$f_D\left(x,a,\tilde{b},c\right)=-\frac{a}{\tilde{b}c}\left(\frac{x}{\tilde{b}}\right)^{-a-1}\left[1+\left(\frac{x}{\tilde{b}}\right)^{-a}\right]^{-1-1/c}.$$

(32)

Therefore one has the relation

$$F_B\left(x,a,\tilde{b},c\right)=1-F_D\left(1/x,a,1/\tilde{b},c\right);\ f_B\left(x,a,\tilde{b},c\right)=-f_D\left(1/x,a,1/\tilde{b},c\right)\left(1/x^2\right)$$
$$\text{and }F_D\left(x,a,\tilde{b},c\right)=1-F_B\left(1/x,a,1/\tilde{b},c\right);\ f_D\left(x,a,\tilde{b},c\right)=-f_B\left(1/x,a,1/\tilde{b},c\right)x^2.$$

(33)

The relation between the Generalized Freundlich-Langmuir function and the BurrXII function can be written using the previous results:

$$\left(\frac{\left(\frac{x}{\tilde{b}}\right)^n}{1+\left(\frac{x}{\tilde{b}}\right)^n}\right)^{\frac{m}{n}}=\left(1+\left(\frac{\tilde{b}}{x}\right)^n\right)^{-\frac{m}{n}}=F_D\left(x,n,\tilde{b},\frac{m}{n}\right)=1-F_B\left(\frac{1}{x},n,\frac{1}{\tilde{b}},\frac{n}{m}\right),\ \tilde{b}=b\left(\frac{m}{n}\right)^{-\frac{1}{n}}.$$

(34)

This allows the GLF isotherm and the Toth [35] isotherm as well as the equivalent Oswin isotherm [36] used in food industry to be part of the $Burr_{XII}$ isotherm family.

The others empirical isotherms [37] [38] correspond to couples of values m and n in the general form (Equation (25)) which give non physical asymptotic behavior and therefore cannot be used over the whole range of concentration or pressure They might give excellent fit over a limited range of data, like the popular Redlich- Peterson isotherm [37] , but cannot give reliable information over the whole sorption process. The same is true for the Freundlich isotherm. In our opinion, as a logical consequence of our work these isotherms should be discarded since we dispose now, with the unified GBS form (Equations (20), (21)), of a four parameter isotherm with a solid theoretical and physical foundation.

We can now derive quite simply the shape of the sorption energy distribution giving rise to the various isotherms we have just derived.

7. SORPTION ENERGY DISTRIBUTIONS

As we already discussed in a previous publication, starting from the thermodynamic relation

$$\varkappa = \exp\left(-e/RT\right) \tag{35}$$

and using the probability theory relation

$$\left|f_p\left(\varkappa\right)\mathrm{d}\varkappa\right| = \left|f_E\left(e\right)\mathrm{d}E\right| \tag{36}$$

it is possible to calculate the sorption energy distribution corresponding to each isotherm. As discussed later, this sorption energy e is the energy which governs the macroscopic thermodynamic properties of the system. It is not the microscopic site energies resulting from the atomic and molecular interactions.

In that way we have obtained the following results:

For the proposed GBS. isotherm derived from the BurrXII distribution function:

$$\phi_{GBS}(e)=\frac{a}{RT}\left(b^{-a}\exp(-ae/RT)\right)\left(1+c\left(b^{-a}\exp(-ae/RT)\right)\right)^{-1-1/c}. \tag{37}$$

The other distributions can be obtained easily:

For $c\rightarrow 0$, we have the distribution corresponding to the BS isotherm

$$\phi_{BS}(e)=\frac{a}{RT}\left(b^{-a}\exp(-ae/RT)\right)\exp\left(b^{-a}\exp(-ae/RT)\right). \tag{38}$$

For $c\rightarrow 1$, we have the distribution corresponding to the Hill-Sips isotherm:

$$\phi_{H}(e)=\frac{a}{RT}\left(b^{-a}\exp(-ae/RT)\right)\left(1+\left(b^{-a}\exp(-ae/RT)\right)\right)^{-2}. \tag{39}$$

It is worth noticing that the BS. distribution has the form of the Gumbel [39] [40] (maximum) extreme value probability distribution function

$$\phi_{G}(e)=\frac{1}{\beta}\mathrm{Exp}\left(-\left(\frac{e-\mu}{\beta}\right)\right)\mathrm{Exp}\left(-\mathrm{Exp}\left(\frac{e-\mu}{\beta}\right)\right) \tag{40}$$

$$\Phi_{G}(E)=\mathrm{Exp}\left(-\mathrm{Exp}\left(-\left(\frac{E-\mu}{\beta}\right)\right)\right) \tag{41}$$

$$\text{with } \beta = RT/a \quad \text{and} \quad \mu = -RT\log b. \tag{42}$$

The standard deviation of this function is well known

$$\beta\pi/\sqrt{6} = (kT/a)\pi/\sqrt{6} \tag{43}$$

confirming the conclusions of reference [1] about the physical signification of the exponent a.

The function $\phi_{GBS}(e)$ corresponding to the new proposed GBS isotherm is one member of the family of generalized Gumbel functions.

$$\phi_{GBS}(e) = \frac{a}{RT}\left(b^{-a}\exp(-ae/RT)\right)\left(1+c\left(b^{-a}\exp(-ae/RT)\right)\right)^{-1-1/c} \tag{44}$$

$$\Phi_{GBS}(E) = \mathrm{Exp}_c\left(-\mathrm{Exp}\left(-\left(\frac{E-\mu}{\beta}\right)\right)\right) \tag{45}$$

$$\text{with } \beta = RT/a \quad \text{and} \quad \mu = -RT\log b \tag{46}$$

It is the symmetric of the Fisher-Tippett [40] [41] generalized extreme value cumulative distribution

$$F_{FT}(E) = \mathrm{Exp}\left(-\mathrm{Exp}_c\left(-\left(\frac{E-\mu}{\beta}\right)\right)\right). \tag{47}$$

It is worth noticing that this last GEV function (Equation (45)) could have been obtained by using the BS isotherm (Equation (12)) and a c-deformed thermodynamic exponential (see Equation (22)) expression

$$\varkappa_i = \mathrm{Exp}_c(-e_i/RT) = \left(1-c(e_i/RT)\right)^{-1/c}. \tag{48}$$

To be complete we have calculated the energy distributions corresponding to the Freundlich-Langmuir isotherm

$$\phi_{FL}(e)=\frac{m}{RT}\frac{b^{-m}\exp(-me/RT)}{\left(1+b^{-n}\exp(-ne/RT)\right)^{\frac{m}{n}+1}}. \tag{49}$$

If $m=n$ (Hill, Sips)

$$\phi_{H}(e)=\frac{n}{RT}\frac{b^{-n}\exp(-ne/RT)}{\left(1+b^{-n}\exp(-nE/RT)\right)^{2}}. \tag{50}$$

If $m=1$ (Toth, Oswin)

$$\phi_{T}(e)=\frac{1}{RT}\frac{b^{-1}\exp(-e/RT)}{\left(1+b^{-n}\exp(-ne/RT)\right)^{\frac{1}{n}+1}}. \tag{51}$$

If $n=1$ (Generalized-Freundlich)

$$f_{GF}(e)=\frac{m}{RT}\frac{b^{-m}\exp(-me/RT)}{\left(1+b^{-1}\exp(-e/RT)\right)^{m+1}}. \tag{52}$$

Some of the these distributions have been obtained earlier by various authors without reference to the pro- bability theory and using the Cerofolini condensation approximation method [21] . Equation (41) was derived in [22] , Equations (39), (50) was derived in [23] and Equation (52) in [24] . They have been used to determine numerically sorption energy distributions from isotherm data and investigate the thermodynamic nature of the sorption from the measured isotherms. The detailed calculations require assumptions on the range of sorption energy, the integrals being performed from $E_{\min}$ to $E_{\max}$ with respect to a reference energy E_0. As a and c tend to 1, one recovers the Langmuir isotherm, the model with a unique sorption energy. Indeed the energy probability density (Equation (39) with $a=1$) is the derivative of a Fermi function and tends to a Heaviside function as T tends to 0. The corresponding pressure or concentration density function has a horizontal asym- ptote at the origin. Physically this means that on a homogeneous surface the pressure range over which sorption takes place (from a few percents to complete coverage) at finite temperature, will be only of one or two order of magnitude, and be narrower as T decreases (and b decreases), an observation already discussed by Roginskii [42] .

8. LANGMUIR, HILL-SIPS AND BROUERS-SOTOLONGO ISOTHERMS OBEY THE MAXIMUM ENTROPY PRINCIPLE

Before concluding this study it is worthwhile to point out that the distribution functions giving the Langmuir, the Hill-Sips-"Langmuir-Freundlich", the Brouers-Sotolongo and Generalized Brouers-Sotolongo can be derived maximizing the Boltzmann-Shannon entropy measure:

$$S(x)=-\int f(x)\log f(x)\mathrm{d}x \tag{53}$$

using the Lagrange multipliers methods [43] [44] with constraints generalizing the ones used for the Weibull distribution [45] by introducing a c-deformation of the power function

$$x_c^a=\frac{1}{c}\ln\left(1+c(x)^a\right) \qquad x_{c->0}^a=x^a \tag{54}$$

in the same spirit as the deformation of the exponential function (Equation (22)). One uses the following con- straints:

$$\int_0^\infty \ln\left(\frac{x}{b}\right)^a f_{GBS}(x;a,b,c)\mathrm{d}x = -\gamma - \ln(c) - \psi(1/c) \tag{55}$$

$$\int_0^\infty \left(\left(\frac{x}{b}\right)^a\right)_c f_{GBS}(x;a,b,c)\mathrm{d}x = \int_0^\infty \frac{1}{c}\ln\left(1+c\left(\frac{x}{b}\right)^a\right) f_{GBS}(x;a,b,c)\mathrm{d}x = 1 \tag{56}$$

where γ is the Euler constant and $\psi(x)$ is the BiGamma function.

For $c->0$ (Brouers-Sotolongo) and $c->1$ (Hill-Sips) and Langmuir $(c=a=1)$ these constraints can be simplified to

$$\int_0^\infty \ln\left(\frac{x}{b}\right) f_{BS}(x;a,b,c)\mathrm{d}x = -\gamma/a \tag{57}$$

$$\int_0^\infty \left(\frac{x}{b}\right)^a f_{BS}(x;a,b)\mathrm{d}x = 1 \tag{58}$$

which are the well known Weibull constraints and

$$\int_0^\infty \ln\left(\frac{x}{b}\right) f_L(x;b)\mathrm{d}x = 0 \tag{59}$$

$$\int_0^\infty \left(\left(\frac{x}{b}\right)^a\right)_{c=1} f_L(x;b)\mathrm{d}x = \int_0^\infty \ln\left(1+\left(\frac{x}{b}\right)^a\right) f_L(x;b)\mathrm{d}x = 1 \tag{60}$$

which are the loglogistic constraints. The fact that these isotherms correspond to the maximum entropy show that they are the best and less biased isotherms when the parameters a, b and c can be determined experi- mentally and therefore should be favoured amongst all the proposed empirical formulas.

9. CONCLUSIONS

In this paper we have shown that a generalized isotherm having the analytical form of a $Burr_{XII}$ cdf is able to generate a whole family of empirical isotherms used in the literature to represent the sorption data of a great number of solid-gas and solid-liquid sorbate-sorbent couples. Due to the fact that the $Burr_{XII}$ and associated functions are used extensively in econometrics, there exists on the market efficient nonlinear fitting computing programs and the use of the GBS isotherm should make obsolete the comparison, often with questionable linear fitting, of experimental isotherms with the various approximations of this more general unified isotherm. Practically since the GBS isotherm interpolates nicely between the BS $(c=0)$ and the Hill-Sips $(c=1)$ isotherm and since the two (a,b) parameters isotherms give generally a reasonably good fit, one can first try both of them and then using these partial results improve the fit with the three (a,b,c) parameters GBS when this is possible given the generally scarcity of data.

The statistical expressions given in the appendix allow a mathematically well defined characterization of the data. Extensions of the $Burr_{XII}$ have been proposed with extra parameters. They belong to the Generalized Beta 2 distribution family and are legitimate cumulative probability functions [46] . Such an extension which might be of interest for huge number of data are irrelevant in sorption problems due to the relatively small number of experimental data.

Another important conclusion of this study is that the energy distributions giving rise to the BS and GBS isotherms belong to the family of extreme value distributions. This is in agreement with the stochastic theory of K. Weron et al. [18] [32] [47] which was developed for relaxation in disordered medias. What matters in highly heterogeneous media is not the detailed microscopic interactions but the extreme value distribution of inter- action energies of dynamically highly correlated mesoscopic clusters (on surfaces, patches, islands). The relation between the phenomenological laws and their microscopic causes has to go through the spatio-temporal scaling properties of these intermediate cooperative regions. This representation allows to average together a large num- ber of extreme probabilistic events to form a predictable picture of the behavior of the entire system. As a con- sequence, the observed tail exponents a and a/c and the analytic form of the equations describing the macro- scopic properties are related to the extreme value cluster energy distributions. The parameter a defined an effective temperature $T^* = T/a$ and $c = q - 1$ is related to the Reyni-Tsallis entropy factor q . In catalysis, the appearance of an effective temperature T^* has been traced to the conditions at which the substrate was prepared and annealed. The active centers regarded as defects once in thermal equilibrium at temperature T^* are "frozen" by sudden cooling (quenching) [42] . More generally an effective temperature $T^* \neq T$ expresses the fact that, due to the frustrations induced by the geometry and the interactions, the couple sorbate-sorbent is not in thermal equilibrium at the experimental temperature T .

Two last remarks have to be made on the range of applicability of the results of this paper. One has to emphasize that it deals with one aspect of sorption i.e. the generalization of the Langmuir isotherm to highly heterogeneous surfaces and solid-liquid interfaces and in some cases of complex composition of sorbates and sorbent. It concerns in particular most works done in water and air decontamination research with pure or treated natural products.

The sorption of simple molecules on smooth surfaces and well defined rough surfaces [48] [49] does not necessarily necessitate an elaborate treatment as used in this paper and the analysis of its isotherms can bring some partial information on the microscopic properties of the surface. In many more complex systems, other phenomena such as wetting, capillarity condensation in pores [50] , as well as diffusion, volume condensation and multi-reactions effects might have to be considered. In those cases, more specific isotherm formulas have to be used [51] . One should also be conscious that the analysis of data with the GBS, Hill-Sips and BS isotherms is relevant only when applied to complete sets of data until saturation.

REFERENCES

1. Brouers, F., Sotolongo-Costa, O., Marquez, F. and Pirard, J.P. (2005) Microporous and Heterogeneous Surface Adsorption Isotherms Arising from Lévy Distributions. Physica A, 349, 271-282.

2. Tien, C. (2008) Remarks on Adsorption Manuscripts Revised and Declined: An Editorial. Journal of Hazardous Materials, 150, 2-3.

3. Ncibi, M.C., Altenor, S., Steffen, M., Brouers, F. and Gaspard, S. (2008) Modelling Simple Compound Adsorption onto Porous and Non-Porous Sorbents Using a Deformed Weibull Exponential Isotherm. Chemical Engineering, 145,

4. Madrakian, T., Afkhami, F. and Ahmadi, M. (2012) Adsorption and Kinetic Studies of Different Organic Dyes onto Magnetic Nanoparticles Loaded Tea Waste and Removal of Them from Wastewater Samples. Spectrochimica Acta A, 99, 102-109.

5. Azizian, S. and Baghery, M. (2014) Enhanced Adsorption of Cu2+ from Aqueous Solution by Ag Doped Nano-Struc- tured ZnO. Journal of Molecular Liquids, 196, 198-203.

6. Vargas, A.M.M., Cazela, A.J., Kunita, M.H. and Silva, T.L. (2011) Adsorption of Methylene Blue on Activated Carbon from Flamboyant Pads: Study of Adsorption Isotherms and Kinetic Models. Chemical Engineering Journal, 168, 722-730.

7. Altenor, S., Carene, B., Enmanuel, E., Lamber, J., Ehrharat, J.J. and Gaspard, S. (2009) Adsorption Studies of Methylene Blue and Phenol onto Vetiver Roots Activated Carbon Prepared by Chemical Activation. Journal of Hazardous Materials, 165, 1029-1039.

8. Hejazifar, M. and Azizian, S. (2012) Adsorption of Cationic and Anionic Dyes onto the Activated Carbon Prepared from Grapevine Rhytidome. Journal of Dispersion Science and Technology, 33, 846-853.

9. Ncibi, M.C., Majoub, B., Steffen, M., Brouers, F. and Gaspard, S. (2009) Sorption Dynamic Investigation of Chromium (VI) onto Posidonia Oceanica Fibres: Kinetic Modelling Using New Generalized Fractal Equation. Biochemical Engineering Journal, 46, 141-146.

10. Altenor, S., Ncibi, M.C., Emmanuel, E. and Gaspard, S. (2012) Textural Characteristic Physiochemical Properties and Adsorption Efficiencies of Caribbean Alga. Turbinaria turbinata and Its Derived Carbonaceous Materials for Water Treatment Application. Biochemical Engineering Journal, 67, 35-44.

11. Hamissa, A.M.B., Brouers, F., Ncibi, M.C. and Steffen, M. (2013) Kinetic Modeling and Methylène Sorption onto Agava Americana Fibres: Fractal Kinetic and Regeneration Studies. Separation Science and Technology, 48, 2834- 2842.

12. Salleh, M., Mahmou, D., Karim, W. and Idris, A. (2011) Cationic and Anionic Dye Adsorption by Agricultural Solid Wastes: A Comprehensive Review. Desalination, 280, 1-13.

13. Figaro, S., Avril, J.P., Brouers, F., Ouensanga, A. and Gaspad, S. (2009) Adsorption Studies of Molasse's Wastewaters on Activated Carbon: Modelling with a New Fractal Kinetic Equation and Evaluation of Kinetic Models. Journal of Hazardous Materials, 161, 649-656.

14. Singh, A., Arutyunov, D., Szymanski, C.M. and Evoy, S. (2012) Bacteriophage Based Probes for Pathogen Detection. Analyst, 137, 3405-3421.

15. Naidoo, R., Singh, A., Arya, S.K., Beadle, B., Glass, N., Tanha, J., Szymanski, C.M. and Evoy, S. (2012) Surface- Immobilization of Chromatographically Purified Bacteriophages for the Optimized Capture of Bacteria. Bacteriophage, 2, 15-24.

16. Singh, A., Poshtiban, S. and Evoy, S. (2013) Recent Advances in Bacteriophage Based Biosensors for Food-Borne Pathogen Detection. Sensors, 13, 1763-1786.

17. Tsallis, C. (2009) Nonadditive Entropy and Nonextensive Statistical Mechanics. An Overview after 20 Years. Brazilian Journal of Physics, 39, 337-357.

18. Brouers, F., Sotolongo-Costa, O. and Weron, K. (2004) Burr, Lévy, Tsallis. Physica A, 344, 409-416.

19. Brouers, F. and Sotolongo-Costa, O. (2005) Relaxation in Heterogeneous Systems: A Rare Events Dominated Phenomenon. Physica A, 356, 359-374.

20. Brouers, F. and Sotolongo-Costa, O. (2006) Generalized Fractal Kinetics in Complex Systems (Application to Biophysics and Biotechnology). Physica A, 368, 165-175.

21. Cerofolini, G.F. (1974) Localized Adsorption on Heterogeneous Surfaces. Thin Solid Films, 23, 129-152.

22. Jaroniec, M. and Piotowska, J. (1986) Isotherm Equations for Adsorption on Heterogeneous Microporous Solids. Monatshefte für Chemie/Chemical Monthly, 117, 7-19.

23. Yaghi, N.Z. (2007) Iron Oxide Based Materials for the Removal of Copper from Drinking Water—A Study of Freundlich Adsorption Isotherms, Site Energy Distributions and Energy Frequency

Distributions. Master's Thesis, Chalmers University of Technology, Göteborg.

24. Kumar, K.V., Monteiro de Castro, M., Martinez-Escandell, M., Molina-Sabio, M. and Rodriguez-Reinoso, F. (2011) A Site Energy Distribution Function from Toth Isotherm for Adsorption of Gases on Heterogeneous Surfaces. Physical Chemistry Chemical Physics, 13, 5753-5759.

25. Burr, I.W. (1942) Cumulative Frequency Functions. The Annals of Mathematical Statistics, 13, 215-232.

26. Maddala, G.S. (1983) Limited-Dependent and Qualitative Variables in Econometrics. Cambridge University Press, Cambridge.

27. Verhulst, P.F. (1845) Recherches mathématiques sur la loi d'accroissement de la population. Nouveaux Mémoires de l'Académie Royale des Sciences et Belles-Lettres de Bruxelles, 18, 14-54.

28. Ausloos, M. and Dirickx, M. (2005) The Logistic Map and the Road to Chaos. Springer, Berlin.

29. Sornette, D. (2004) Critical Phenomena in Natural Sciences. 2nd Edition, Springer, Berlin.

30. Jovanovic, D.S. (1969) Physical Adsorption of Gazes. Kolloid-Zeitschrift und Zeitschrift für Polymere, 235, 1214- 1225.

31. Monod, J., Wyman, J. and Changeux, J.P. (1965) On the Nature of Allosteric Transitions: A Plausible Model. Journal of Molecular Biology, 12, 88-118.

32. Stanislavsky, A. and Weron, K. (2013) Is There a Motivation for a Universal Behaviour in Molecular Populations Undergoing Chemical Reactions? Physical Chemistry Chemical Physics, 15, 15595-15601.

33. Marczewski, A.W. and Jaroniec, M. (1983) A New Isotherm Equation for Single-Solute Adsorption from Dilute Solutions on Energetically Heterogeneous Solids. Monatshefte für Chemie/Chemical Monthly, 114, 711-715.

34. Dagum, C. (1977) A New Model of Personal Income-Distribution-Specification and Estimation. Economie Appliquée, 30, 413-437.

35. Toth, J. (1995) Uniform Interpretation of Gas/Solid Adsorption. Advances in Colloid and Interface Science, 55, 1-239.

36. Oswin, C.R. (1946) The Kinetics of Package Life III. The Isotherm. Journal of the Society of Chemical Industry, 65, 419-421.

37. Redlich, O. and Peterson, D.L. (1959) A Useful Adsorption Isotherm. Journal of Physical Chemistry, 63, 1024.

38. Müller, G., Radke, C.J. and Prausnitz, J.M. (1985) Adsorption of Weak Organic Electrolytes from Dilute Aqueous Solution onto Activated Carbon. Part I. Single-Solute Systems. Journal of Colloid and Interface Science, 103, 466-483.

39. Gumbel, E.J. and Lieblein, J. (1954) Statistical Theory of Extreme Values and Some Practical Applications: A Series of Lectures (Vol. 33). US Government Printing Office, Washington.

40. Coles, S. (2001) An Introduction to Statistical Modeling of Extreme Values. Springer Verlag, Berlin.

41. Fischer, R.A. and Tippett, L.H.C. (1928) Limiting Forms of the Frequency Distribution of the Largest or Smallest Member of a Sample. Mathematical Proceedings of the Cambridge Philosophical Society, 24, 180-190.

42. Young, D.M. and Crowell, A.D. (1962) Physical Adsoption of Gases, Chapter 7, Buttenwerks, London.

43. Papalexiou, S.M. and Koutsoyiannis, D. (2012) Entropy Based Derivation of Probability Distributions: A Case Study to Daily Rainfall. Advances in Water Resources, 45, 51-57.

44. Park, S.Y. and Bera, A.K. (2009) Maximum Entropy Autoregressive Conditional Heteroskedasticity Model. Journal of Econometrics, 150, 219-230.

45. Kapur, J.N. (1993) Maximum Entropy Models in Science and Technology. Wiley Eastern Limited.

46. McDonald, J.B. (1984) Some Generalized Functions for the Size Distribution of Income. Journal of the Econometric Society, 52, 647-663.

47. Hetman, P. and Weron, K. (2004) Extreme-Value Approach to the Tsallis' Superstatistics. Acta Physica Polonica B, 35, 1375-1386.

48. Sahouli, B., Blacher, S. and Brouers, F. (1997) Applicability of the Fractal FHH Equation. Langmuir, 13, 4391-4394.

49. Blacher, S., Pirard, R., Pirard, J.P., Sahouli, B. and Brouers, F. (1997) On the Texture Characterization of Mixed SiO2-ZrO2 Aerogels Using the Nitrogen Adsorption-Desorption Isotherms: Classical and Fractal Methods. Langmuir, 13, 1145-1149.

50. Sahouli, B., Blacher, S. and Brouers, F. (1996) Fractal Surface Analysis by Using Nitrogen Adsorption Data: The Case of the Capillary Condensation Regime. Langmuir, 12, 2872-2874.

51. Dabrowski, A. (2001) Adsorption: From Theory to Practice. Advances in Colloid and Interface Science, 93, 135-224.

52. Brouers, F. (2013) Sorption Isotherms and Probability Theory of Complex Systems. arXiv:1309.5340.

Chapter 4

Sampling Designs with Linear and Quadratic Probability Functions

Lennart Bondesson[1], Anton Grafström[2], Imbi Traat[3]

[1]Department of Mathematics and Mathematical Statistics, Umeå University, Umeå, Sweden

[2]Department of Forest Resource Management, Swedish University of Agricultural Sciences, Umeå, Sweden

[3]Institute of Mathematical Statistics, University of Tartu, Tartu, Estonia

ABSTRACT

Fixed size without replacement sampling designs with probability functions that are linear or quadratic functions of the sampling indicators are defined and studied. Generality, simplicity, remarkable properties, and also somewhat restricted flexibility characterize these designs. It is shown that the families of linear and quadratic designs are closed with respect to sample complements and with respect to conditioning on sampling outcomes for specific units.

Relations between inclusion probabilities and parameters of the probability functions are derived and sampling procedures are given.

KEYWORDS

Complementary Midzuno Design; Conditional Sample; Inclusion Probability; Midzuno Design; Mixture of Designs; Parameters of Design; Sample Complement; Sinha Design

1. INTRODUCTION

In the first part of the paper, we consider the most general fixed size without replacement sampling design with a probability function that is linear in the sampling inclusion indicators—the linear sampling design. The linear design, being an unequal probability design, is remarkable due to simple explicit relations between inclusion probabilities and parameters of the probability function. This enables sampling with desired inclusion probabilities, design-unbiased estimation and variance estimation. As special cases, the design covers the classical Midzuno [1] and the complementary Midzuno (see [2]) designs. It is shown that the linear design can be seen as a mixture of the two types of Midzuno designs. It is also shown that the family of linear designs is closed with respect to conditioning on sampling outcomes. This property, as well as the mixture representation, offers easy methods for sampling from the linear design. The family is also closed with respect to sample complements, i.e. the complement of a sample from a linear design is a sample from another linear design.

In the second part of the paper, the fixed size without replacement design with quadratic probability function—the quadratic sampling design—is defined and studied. It is the natural extension of the linear design. A classical design by Sinha [3] is a special case of the quadratic design. His design aimed at sampling with prescribed second-order inclusion probabilities. Together with Sinha's design, two other special quadratic designs are studied more closely. These three designs are easy to use for sampling. There are explicit expressions for the second-order inclusion probabilities of Sinha's design. In the general case, the formulas are more complicated. A lemma is proved which relates the second-order inclusion probabilities and the parameters of the quadratic design. Like the family of linear designs, the family of quadratic designs is closed with respect to sample complements and with respect to conditioning on sampling outcomes. The last property makes list-sequential sampling from these designs efficient.

The linear and the quadratic designs are simple but somewhat restricted when the aim is sampling with prescribed firstor second-order inclusion probabilities. They cannot be used for all such probabilities. In the final section of the paper, possible extensions are mentioned.

2. LINEAR SAMPLING DESIGNS

We treat without replacement (WOR) sampling designs of size n from a population of size N. Let $\boldsymbol{I}=(I_1,I_2,\cdots,I_N)$ be the binary random sampling inclusion vector. A sampling design is given by its probability function $p(\boldsymbol{x})=\Pr(\boldsymbol{I}=\boldsymbol{x})$. Let $S_n=\left\{\boldsymbol{x}\in\{0,1\}^N;\ |\boldsymbol{x}|=\sum_{k=1}^N x_k=n\right\}$, i.e. S_n is the set of all possible "samples" $\boldsymbol{x}$ of size n.

Definition 1. A sampling design of size $n\,(1\le n\le N-1)$ is called linear if there are real constants $c_1,c_2,\ldots,c_N$ such that $p(\boldsymbol{x})\propto\sum_{k=1}^N c_k x_k,\ \boldsymbol{x}\in S_n$.

Of course, equal cks give SRSWOR. Below it is always assumed that the cks are normalized to have sum 1.

Then $p(\boldsymbol{x})=C\sum_{k=1}^N c_k x_k$, where $C=1\Big/\binom{N-1}{n-1}$. In fact,

$$1=\sum_{x\in S_n} p(x)=C\sum_{k=1}^{N} c_k \sum_{x\in S_n} x_k = C\sum_{k=1}^{N} c_k \binom{N-1}{n-1} = C\binom{N-1}{n-1}.$$

The inclusion probabilities $\pi_i = \Pr(I_i = 1)$ of the linear design are given by

$$\pi_i = c_i + (1-c_i)\frac{n-1}{N-1}. \tag{1}$$

Indeed, since $\sum_{k=1}^{N} c_k = 1,$

$$\pi_i = \sum_{x\in S_n} x_i p(x) = C\sum_{k=1}^{N} c_k \sum_{x\in S_n} x_i x_k = C\left(c_i \binom{N-1}{n-1} + (1-c_i)\binom{N-2}{n-2}\right).$$

Since $0 \le \pi_i \le 1,$ (1) shows that we must have $-\frac{n-1}{N-n} \le c_i \le 1.$ The c_i s can be expressed in terms of the π_i s as

$$c_i = \frac{N-1}{N-n}\left(\pi_i - \frac{n-1}{N-1}\right). \tag{2}$$

By similar algebra, the second-order inclusion probabilities $\pi_{ij} = \Pr(I_i I_j = 1)$ are given by

$$\pi_{ij} = \frac{n-1}{N-1}(c_i + c_j) + (1 - c_i - c_j)\frac{(n-1)(n-2)}{(N-1)(N-2)}, \quad i \ne j,$$

and hence

$$\pi_{ij} = \frac{n-1}{N-2}\left(\pi_i + \pi_j - \frac{n}{N-1}\right), \quad i \ne j, \tag{3}$$

i.e. the π_{ij} s are linear in the first-order inclusion probabilities.

Without restriction we may assume that the cks are given in increasing order. Obviously, in order to get $p(\boldsymbol{x}) \geq 0,$ it is then necessary and sufficient that $\sum_{k=1}^{n} c_k \geq 0.$ We obtain the following theorem.

Theorem 1. Let $0 \leq \pi_1 \leq \pi_2 \leq \cdots \leq \pi_N \leq 1$ be given numbers with sum n. Then there is a linear sampling design with these numbers as inclusion probabilities if and only if $\frac{1}{n}\sum_{k=1}^{n} \pi_k \geq \frac{n-1}{N-1}.$

Proof. By the relation (2), we see that $\sum_{k=1}^{n} c_k \geq 0 \Leftrightarrow \frac{1}{n}\sum_{k=1}^{n} \pi_k \geq \frac{n-1}{N-1}.$ Ω

To sample from a linear sampling design, we may use the well-known acceptance-rejection (AR) technique.

A constant A such that $p(\boldsymbol{x}) = C\sum_{k=1}^{N} c_k x_k \leq A p_{SRS}(\boldsymbol{x}) = A \Big/ \binom{N}{n}$ for all $\boldsymbol{x} \in S_n$ is first found. Assuming that the cks are in increasing order, it suffices to put $A = \frac{N}{n}\sum_{k=N-n+1}^{N} c_k.$ Then we generate a tentative sample x according to SRSWOR of size n and a random number $U \sim U(0,1).$ The sample is accepted as a sample from the linear design if

$$U \leq \frac{C\sum_{k=1}^{N} c_k x_k}{A p_{SRS}(\boldsymbol{x})} = \frac{N}{n}\frac{\sum_{k=1}^{N} c_k x_k}{A} = \frac{\sum_{k=1}^{N} c_k x_k}{\sum_{k=N-n+1}^{N} c_k}.$$

The full procedure is repeated until a sample is accepted. The acceptance rate equals $\sum_{k=1}^{N} c_k \pi_k \Big/ \sum_{k=N-n+1}^{N} c_k$. To be seen later on, also other sampling techniques exist.

There are two special linear sampling designs: the Midzuno design and the complementary Midzuno design.

The Midzuno design has the probability function $p_M(\boldsymbol{x}) \propto \sum_{k=1}^{N} a_k x_k$, $\boldsymbol{x} \in S_n$, with $\sum_{k=1}^{N} a_k = 1$ and $a_k \geq 0$ for all k. Apparently it is a linear design. We may sample from the design by selecting one unit according to the probabilities a_k and then $n-1$ further units according to SRSWOR from the remaining $N-1$ ones. The design was introduced in [1] (considering the a_k s as proportional to auxiliary variables). Tillé ([4] , p. 117) generalizes it and gives many further references concerning the design. Brewer and Hanif ([5] , p. 25) remark that the inequalities on the inclusion probabilities are restrictive.

The complementary Midzuno design (see [2]) has the probability function

$$p_{cM}(\boldsymbol{x}) = \binom{N-1}{n}^{-1} \sum_{k=1}^{N} b_k (1 - x_k), \ \boldsymbol{x} \in S_n,$$

with $\sum_{k=1}^{N} b_k = 1$ and $b_k \geq 0$ for all k. This design is considered in [6] with the b_k s proportional to auxiliary variables. We may sample from the design by removing one unit according to the probabilities b_k and sample n units by SRSWOR among the remaining $N-1$ ones. For the complementary Midzuno design, we then get

$\pi_i = \frac{n}{N-1}(1-b_i).$ The formula for the second-order inclusion probabilities coincides with (3). Since

$$\sum_{k=1}^{N} b_k (1-x_k) = \sum_{k=1}^{N} \left(n^{-1} - b_k\right) x_k,\ \ \boldsymbol{x} \in S_n,$$

we also see that the complementary Midzuno design is a linear design with coefficients $\&\wp_k = \frac{n}{N-n}\left(\frac{1}{n} - b_k\right)$ with sum 1. If $b_k \le 1/n$ for all k, it is even a Midzuno design and hence the two designs overlap each other.

Remark 1. Let $0 \le \pi_1 \le \pi_2 \le \cdots \le \pi_N \le 1$ have sum n, i.e. $\sum_{k=1}^{N} \pi_k = n.$ Then, by (2), these numbers are inclusion probabilities for a Midzuno design iff (a) $\pi_1 \ge \frac{n-1}{N-1}.$ Obviously (a) implies the less restrictive inequality in Theorem 1. The π_i s are inclusion probabilities of a complementary Midzuno design iff (b) $\pi_N \le \frac{n}{N-1}.$ Of course, also (b) implies the inequality in Theorem 1. In fact, assuming the contrary that $\frac{1}{n}\sum_{k=1}^{n} \pi_k < \frac{n-1}{N-1},$ we get by (b) that $\sum_{k=1}^{N} \pi_k < \frac{n(n-1)}{N-1} + \frac{(N-n)n}{N-1} = n,$ which is a contradiction as the sample size is n.

The family of linear designs can be considered as closed with respect to sample complements. More precisely, the complement of any sample of size n from a linear design, is a sample of size $N-n$ from another linear design. This follows from the relation

$$\sum_{k=1}^{N} c_k x_k = \sum_{k=1}^{N}\left(\frac{1}{N-n} - c_k\right)(1-x_k), \quad \boldsymbol{x} \in S_n.$$

Another interesting property of the family of linear designs is that it is closed with respect to conditioning on the sampling outcomes for specific units. For instance, if we know that unit 1 is selected (i.e. $x_1 = 1$), then the probability function for the remaining size $n-1$ sampling among $N-1$ units is still linear. In fact, given that $\pi_1 > 0$ and that $I_1 = x_1 = 1,$ we have by (1) that $c_1 > -\frac{n-1}{N-n}.$ The conditional probability function of $\boldsymbol{I}_2 = (I_2, I_3, \ldots, I_N)$ is then given by

$$\Pr(\boldsymbol{I}_2 = \boldsymbol{x}_2) \propto \frac{1}{\pi_1}\left(\sum_{k=2}^{N} c_k x_k + c_1\right) \propto \sum_{k=2}^{N}\left(c_k + c_1\frac{1}{n-1}\right)x_k, \quad |\boldsymbol{x}_2| = n-1.$$

The new coefficients $c_k + \frac{c_1}{n-1}, k \geq 2,$ have sum $1 - c_1 + \frac{N-1}{n-1}c_1 = 1 + \frac{N-n}{n-1}c_1 > 0$ by the inequality for c_1 above. Normalizing them, we get coefficients $c_k^1 = \frac{(n-1)c_k + c_1}{(N-n)c_1 + n - 1}$ with sum 1. If it is known that unit 1 is not selected ($x_1 = 0$ and $\pi_1 < 1,\ c_1 < 1$), the new coefficients are simply given by $c_k^0 = c_k \Big/ \sum_{i=2}^{N} c_i\,, k \geq 2.$

It follows that samples from a linear design can easily be generated list-sequentially. The first unit in the population is sampled with probability $\pi_1 = c_1 + (1-c_1)\frac{n-1}{N-1}.$ If it is selected, then the second unit is sampled with probability $c_2^1 + \left(1 - c_2^1\right)\frac{n-2}{N-2}$. Else the second unit is sampled with probability $c_2^0 + \left(1 - c_2^0\right)\frac{n-1}{N-2}$ etc.

3. A MIXTURE RESULT

A linear design can also be called a mixed Midzuno design because of the following theorem.

Theorem 2. A linear sampling design $p(x) \propto \sum_{k=1}^{N} c_k x_k, x \in S_n$, where $\sum_{k=1}^{N} c_k = 1$, is a mixture of a Midzuno design and a complementary Midzuno design. It is a Midzuno design if $c_k \geq 0$ for all k and it is a complementary one if $c_k \leq 1/(N-n)$ for all k.

The components of the mixture are not unique. The very last statement in the theorem follows from the re-writing $\sum_{k=1}^{N} c_k x_k = \sum_{k=1}^{N}\left(\frac{1}{N-n} - c_k\right)\left(1 - x_k\right), x \in S_n$. The full proof of the theorem is somewhat technical and is given in an appendix. The proof yields the following procedure for finding suitable components of the mixture. It is assumed that the design is not a pure Midzuno design and that the units in the population are ordered such that $c_1 \leq c_2 \leq \cdots \leq c_m < 0 \leq c_{m+1} \leq \cdots \leq c_N$, where $1 \leq m < n$ (since $\sum_{k=1}^{n} c_k \geq 0$).

Procedure:

Let $\nu = m$. While $c_{\nu+1} < -\frac{1}{n-\nu}\sum_{k=1}^{\nu} c_k$, let $\nu = \nu + 1$.

(As $\sum_{k=1}^{n} c_k \geq 0$, the final ν is less than n.)

Then put $\beta = -\frac{N-n}{n-\nu}\sum_{k=1}^{\nu} c_k$, $\alpha = 1 - \beta$, and $\gamma = \beta\frac{n}{N-n}$.

Finally, let the parameters for the components be defined by

$$a_k = \begin{cases} 0, & k = 1,2,\ldots,\nu \\ \left(c_k - \frac{\gamma}{n}\right)\Big/\alpha, & k = \nu+1,\ldots,N \end{cases} \quad \text{and} \quad b_k = \begin{cases} \frac{1}{n} - \frac{c_k}{\gamma}, & k = 1,2,\ldots,\nu \\ 0, & k = \nu+1,\ldots,N. \end{cases}$$

The first component, the Midzuno one, is used with probability α and the second component, the complementary Midzuno one, is used with probability β.

Example 1. Let $N = 8$, $n = 6$, and $\boldsymbol{c} = \left[-\frac{1}{4}, -\frac{1}{10}, -\frac{1}{20}, \frac{1}{10}, \frac{1}{5}, \frac{1}{4}, \frac{7}{20}, \frac{1}{2}\right]$. Obviously $m = 3$. It is not possible to use $\nu = 3$ since $c_4 = \frac{1}{10} < -\frac{1}{n-3}\sum_{k=1}^{3} c_k = \frac{8}{60}$. The procedure gives us $\nu = 4$, $\beta = \frac{3}{10}$, $\alpha = \frac{7}{10}$ and $\gamma = \frac{9}{10}$ and $\boldsymbol{a} = \left[0,0,0,0,\frac{1}{14},\frac{1}{7},\frac{2}{7},\frac{1}{2}\right]$ and $\boldsymbol{b} = \left[\frac{4}{9},\frac{5}{18},\frac{2}{9},\frac{1}{18},0,0,0,0\right]$. However, since

$c_k \leq \frac{1}{N-n} = \frac{1}{2}$, the design is also a pure complementary Midzuno design with $b_k = \frac{1}{n} - \frac{N-n}{n} c_k$ and hence $\boldsymbol{b} = \left[\frac{1}{4}, \frac{1}{5}, \frac{11}{60}, \frac{2}{15}, \frac{1}{10}, \frac{1}{12}, \frac{1}{20}, 0\right]$.

Theorem 2 implies another simple way of generating samples from the linear design. First it is decided by a random choice whether a Midzuno or a complementary Midzuno design should be used. Then one of these designs is applied. In Tillé's ([4] , pp. 99-104) terminology, we can see this technique as a special splitting technique which after two steps ends with SRSWOR.

4. QUADRATIC SAMPLING DESIGNS

There is a natural extension of the linear designs. In this section we look at fixed size designs with quadratic probability functions. We could use ordinary quadratic forms. However, it is more appropriate to sum over sets of size 2 than over pairs with ordered elements. Below $\sum_{\{i,j\}}$ means a sum over the $\binom{N}{2}$ sets $\{i,j\}$, $j \neq i$.

Definition 2. A sampling design of size $n(2 \leq n \leq N-2)$ is called quadratic if there are real constants d_{ij} with $d_{ji} = d_{ij}$ and $\sum_{\{i,j\}} d_{ij} = 1$ such that $p(\boldsymbol{x}) \propto \sum_{\{i,j\}} d_{ij} x_i x_j, \boldsymbol{x} \in S_n.$

In particular, if $n = 2$ then d_{ij} is the probability to select the sample $\{i,j\}$. The normalization constant can be shown to be equal to the reciprocal of $\binom{N-2}{n-2}$. It can also be shown that the parameters (d_{ij}) are uniquely determined by the design. To get $p(\boldsymbol{x}) \geq 0$ for all $\boldsymbol{x} \in S_n$, it is necessary and sufficient that $\sum_{\{i,j\}\subseteq A} d_{ij} \geq 0$ for all subsets A of $\{1,2,\cdots,N\}$ of size n. This means that the possible parameters (d_{ij}) form a convex set in R^M, where $M=\binom{N}{2}$. However, this set is rather complicated if n is not very small.

For $2 \leq n \leq N-2$, a linear design can also be seen as a quadratic design, because $(n-1)x_i = \sum_{j;j\neq i} x_i x_j$, $\boldsymbol{x} \in S_n$. There are three natural quadratic designs with their basic parameters symmetric and nonnegative:

(a) $p_a(\boldsymbol{x}) \propto \sum_{\{i,j\}} a_{ij} x_i x_j, \ \boldsymbol{x} \in S_n.$

(b) $p_b(\boldsymbol{x}) \propto \sum_{\{i,j\}} b_{ij}(1-x_i)(1-x_j), \ \boldsymbol{x} \in S_n.$

(c) $p_c(\boldsymbol{x}) \propto \sum_{\{i,j\}} c_{ij} \dfrac{x_i(1-x_j)+x_j(1-x_i)}{2}, \ \boldsymbol{x} \in S_n.$

Assuming that $\sum_{\{i,j\}} a_{ij} = 1$, $\sum_{\{i,j\}} b_{ij} = 1$, and $\sum_{\{i,j\}} c_{ij} = 1$, we readily find that for the three cases the normalization constants are the reciprocals of $\binom{N-2}{n-2}$, $\binom{N-2}{n}$, and $\binom{N-2}{n-1}$, respectively. It is not clear that the designs (b) and (c) are quadratic according to Definition 2 but it will be obvious later on. It will also follow that the complement of a sample from a quadratic design of size n can be seen as a sample from a quadratic design of size $N-n$.

The design (a) corresponds to selecting one pair of units according to the probabilities a_{ij} and then $n-2$ further ones by SRSWOR. The design (b) was considered by Sinha in [3] . Remove one pair of units according to the probabilities b_{ij} with sum 1 and then choose n units by SRSWOR among the remaining ones. The design (c) corresponds to selecting one pair according to the probabilities c_{ij} and then keeping one of the units for the sample and removing the other one. Then $n-1$ units are selected by SRSWOR among the $N-2$ non-selected units.

Remark 2. An extension of the design (c) would be to let $p_c(\boldsymbol{x}) \propto \sum_{i,j} c_{ij} x_i (1-x_j),\ \boldsymbol{x} \in S_n,$ with $c_{ij} \geq 0$ and $c_{ii} = 0$ but not necessarily $c_{ij} = c_{ji}$. In this case the units i and j have different roles. In the symmetric case the new parameters c_{ij} equal the previous ones multiplied by 1/2. This extension is not considered further here.

Using the facts that for $\boldsymbol{x} \in S_n,\ (n-1)x_i = \sum_{j;j\neq i} x_i x_j,$ and $\binom{n}{2} = \sum_{\{i,j\}} x_i x_j$ and

$$\sum_{\{i,j\}} b_{ij}\left(x_i + x_j\right) = \sum_{\{i,j\}} \frac{B_i + B_j}{n-1} x_i x_j, \text{ where } B_i = \sum_{j;j\neq i} b_{ij},$$

it is not difficult to see that also the designs (b) and (c) are indeed quadratic designs according to Definition 2 with parameters d_{ij} that are not necessarily nonnegative. More explicitly we have, with

$$B = \sum_{i=1}^{N} B_i = 2\sum_{\{i,j\}} b_{ij} = 2,$$

$$\sum_{\{i,j\}} b_{ij}\left(1-x_i\right)\left(1-x_j\right) = \sum_{\{i,j\}} \left(b_{ij} - \frac{B_i + B_j}{n-1} + \frac{B}{n(n-1)} \right) x_i x_j,$$

$$\sum_{\{i,j\}} c_{ij} \frac{x_i(1-x_j)+x_j(1-x_i)}{2} = \sum_{\{i,j\}} \left(\frac{C_i+C_j}{2(n-1)} - c_{ij} \right) x_i x_j, \text{ where } C_i = \sum_{j;j\neq i} c_{ij}.$$

One can then realize that $p_b(\boldsymbol{x}) \propto \sum_{\{i,j\}} d_{ij}^b x_i x_j,$ where

$$d_{ij}^b = \frac{n(n-1)}{(N-n)(N-n-1)} \left(b_{ij} - \frac{B_i+B_j}{n-1} + \frac{B}{n(n-1)} \right) \text{ and } \sum_{\{i,j\}} d_{ij}^b = 1. \tag{4}$$

Further

$$p_c(\boldsymbol{x}) \propto \sum_{\{i,j\}} d_{ij}^c x_i x_j, \text{ where } d_{ij}^c = \frac{n-1}{N-n} \left(\frac{C_i+C_j}{2(n-1)} - c_{ij} \right) \text{ and } \sum_{\{i,j\}} d_{ij}^c = 1.$$

Given parameters d_{ij} with sum 1 we may solve the equations $d_{ij}^b = d_{ij}$ for possible nonnegative solutions b_{ij} and the equations $d_{ij}^c = d_{ij}$ for possible nonnegative solutions c_{ij}. In fact, we have

$$b_{ij} = \frac{(N-n)(N-n-1)}{n(n-1)} \left(d_{ij} - \frac{D_i+D_j}{N-n-1} + \frac{D}{(N-n)(N-n-1)} \right), \tag{5}$$

where $D_i = \sum_{j;j\neq i} d_{ij}$ and $D = \sum_{i=1}^{N} D_i = 2$. This result is a consequence of formula (4) but with the sample of size n replaced by its complement of size $N-n$. Somewhat more complicated calculations show that, at least if $2n \neq N$,

$$c_{ij} = \frac{N-n}{(N-2n)(n-1)} (D_i + D_j) - \frac{D}{(N-2n)(n-1)} - \frac{N-n}{n-1} d_{ij}. \tag{6}$$

This gives us the following theorem.

Theorem 3. A quadratic sampling design $p(\boldsymbol{x}) \propto \sum_{\{i,j\}} d_{ij} x_i x_j,\ \boldsymbol{x} \in S_n,$ with $\sum_{\{i,j\}} d_{ij} = 1$ is a design of type (a) iff $d_{ij} \geq 0.$ It is design of type (b) (Sinha design) iff

$$d_{ij} - \frac{D_i + D_j}{N-n-1} + \frac{D}{(N-n)(N-n-1)} \geq 0.$$

It is a design of type (c) iff

$$\frac{1}{N-2n}\left(D_i + D_j\right) - \frac{D}{(N-2n)(N-n)} - d_{ij} \geq 0.$$

It can happen that a design is of all the three types simultaneously. For example, SRSWOR is such a design.

Example 2. Let $N = 5, n = 3$ and let the design be defined by

$$d_{12} = \frac{1}{6}, d_{13} = \frac{1}{12}, d_{14} = -\frac{1}{12}, d_{15} = \frac{1}{6}, d_{23} = d_{24} = \frac{1}{4}, d_{25} = d_{34} = d_{35} = 0, d_{45} = \frac{1}{6}.$$

By (5) and (6), the corresponding band c-parameters are then given by

$$b_{12} = \frac{1}{18}, b_{13} = \frac{5}{36}, b_{14} = \frac{1}{12}, b_{15} = \frac{1}{6}, b_{23} = b_{24} = \frac{1}{12}, b_{25} = 0, b_{34} = b_{35} = \frac{1}{9}, b_{45} = \frac{1}{6}$$

$$c_{12} = -\frac{1}{6}, c_{13} = \frac{1}{4}, c_{14} = \frac{5}{12}, c_{15} = \frac{1}{6}, c_{23} = c_{24} = -\frac{1}{4}, c_{25} = 0, c_{34} = c_{35} = \frac{1}{3}, c_{45} = \frac{1}{6}.$$

Thus in this case the design is not of type (a) or (c) but of type (b).

In the linear case we saw that a general linear design can be represented as a mixture of a Midzuno and a complementary Midzuno design. It would be desirable that a general quadratic design could be represented as a mixture of the three designs (a), (b), and (c) (or its extension). However, such a result is not likely to hold.

For the design (b) the second-order inclusion probabilities π_{ij} (which determine the first-order ones since $(n-1)\pi_i = \sum_{j;j\neq i} \pi_{ij}$) are easy to find. We have

$$\pi_{ij} = \frac{n(n-1)}{(N-2)(N-3)} \sum_{\{k,l\};\{k,l\}\cap\{i,j\}=\varnothing} b_{kl}.$$

It is also easy to find the b-parameters that correspond to (π_{ij}). In fact,

$$b_{ij} = \frac{(N-2)(N-3)}{n(n-1)}\pi_{ij} - \frac{N-2}{n}\left(\pi_i + \pi_j\right) + 1.$$

This result is due to Sinha [3] but more generally also a special case of Lemma 1 below. Sinha used his result to find a method to sample with prescribed second-order inclusion probabilities. However, for the method to work with nonnegative b_{ij}, strong restrictions on the π_{ij} s are needed.

Lemma 1. Let $q_{ij}, i \neq j,$ be given numbers with $q_{ij} = q_{ji}$. Then the equations

$$\sum_{\{k,l\};\{k,l\}\cap\{i,j\}=\varnothing} c_{kl} = q_{ij}, \quad i \neq j, \tag{7}$$

in the unknowns $c_{ij} = c_{ji}, i \neq j,$ have the solution

$$c_{ij} = q_{ij} - \frac{Q_i + Q_j}{N-3} + \frac{Q}{(N-2)(N-3)}, \text{ where } Q_i = \sum_{j;j\neq i} q_{ij}, Q = \sum_{i=1}^{N} Q_i. \tag{8}$$

We only have to put $q_{ij} = \pi_{ij}\frac{(N-2)(N-3)}{n(n-1)}$, to get the earlier solution $b_{ij}\left(=c_{ij}\right)$ from the lemma. The proof of Lemma 1 is given in the appendix.

For a design given by $p(\boldsymbol{x}) \propto \sum_{\{i,j\}} d_{ij}x_i x_j$, the formulas for the second-order inclusion probabilities π_{ij} are somewhat complicated. However, we can first find b_{ij} from formula (5) and then use the π_{ij} formula above. The inverse procedure $(\pi \rightarrow b \rightarrow d)$ can also be used to find d-parameters corresponding to given second-order inclusion probabilities. The “mixed case” (c) can be handled in the same way. A computer program makes these procedures simple to use.

To sample from the general quadratic design, we could use an AR-technique by dominating $p(\boldsymbol{x})$ by a multiple of $p_{SRS}(\boldsymbol{x})$. However, it is probably simpler to use list-sequential sampling. In fact, given $I_1 = x_1$ (0 or 1), the probability function for the remaining sample of size n or $n-1$ is again quadratic, cf. Section 2. But we must recalculate parameters (coefficients) and calculate first-order inclusion probabilities. The first-order inclusion probability π_1 is given by (assuming that the d-sum is 1)

$$\pi_1 = \sum_{l;l\geq 2} d_{1l} + \frac{n-2}{N-2} \sum_{\{k,l\},l>k\geq 2} d_{kl}.$$

Given that $I_1 = x_1$ with $x_1 = 0$, the new d-coefficients (with sum 1) are just

$$d_{ij}^0 = \frac{d_{ij}}{\sum_{\{k,l\},l>k\geq 2} d_{kl}}, \quad i,j \geq 2, i \neq j.$$

If $x_1 = 1$ and $n > 2$, the new d-coefficients (with sum 1) are instead given by

$$d_{ij}^1 \propto d_{ij} + \frac{d_{1i} + d_{1j}}{n-2}, \quad i,j \geq 2, i \neq j.$$

For $n = 2$ the conditional design is a linear design with parameters proportional to d_{1i}.

5. ADDITIONAL COMMENTS

As recently becoming more common in sampling articles, we have used sampling indicators and focused on the probability function of these. The names linear design and quadratic design become very natural for this reason.

There are several advantages of the linear design: there are simple relations between the parameters and the inclusion probabilities of first and second order. It is also easy to sample. Thus the design is easy to use and for the Horvitz-Thompson estimate of a population total, we can also readily find a variance estimate. The drawback of the design is its lack of sufficient flexibility: although much more general than SRSWOR, it is not able to cover all possible first-order inclusion probabilities.

The quadratic design is more complicated than the linear one. By using a quadratic design, it is possible to sample with prescribed second-order inclusion probabilities. However, it cannot be used for all such possible inclusion probabilities and therefore this design is also not flexible enough.

To get more flexibility, the linear and quadratic functions need to be complemented by binomial factors. To sample with prescribed first-order inclusion probabilities, we may generally use a probability function of the form

$$p(\boldsymbol{x}) \propto \prod_{i=1}^{N} p_i^{x_i} (1-p_i)^{1-x_i} \times \sum_{k=1}^{N} c_k x_k, \quad \boldsymbol{x} \in S_n,$$

where the parameters are suitably chosen. At least three well-known designs are of this form (with nonnegative c_k): the conditional Poisson design, the Sampford design, and the Pareto design, cf. [7] . For these three cases, the linear function $\sum_{k=1}^{N} c_k x_k$ is proportional to $\sum_{k=1}^{N} b_k (1-x_k)$ with nonnegative parameters b_k.

To sample with prescribed second-order inclusion probabilities, we may use a design with

$$p(\boldsymbol{x}) \propto \prod_{i=1}^{N} p_i^{x_i} (1-p_i)^{1-x_i} \times \sum_{\{k,l\}} b_{kl} (1-x_k)(1-x_l), \quad \boldsymbol{x} \in S_n.$$

Letting $p_i = \pi_i$ (the desired first-order inclusion probabilities) and using Lemma 1, it is possible to calculate the suitable parameters b_{kl} without much effort. Details will be presented in a planned forthcoming paper.

This design is much more flexible than the quadratic design but it has not full flexibility. A fully flexible design uses a probability function of the form $p(x) \propto \exp\left(\sum_{\{i,j\}} a_{ij} x_i x_j\right), x \in S_n,$ but it is not easy to find the parameters [8] .

ACKNOWLEDGEMENTS

This research was supported by the Estonian Science Foundation grant 8789.

REFERENCES

1. Midzuno, H. (1952) On the Sampling System with Probability Proportional to Sum of Sizes. Annals of the Institute of Statistical Mathematics, 3, 99-107.

2. Bondesson, L. and Traat, I. (2013) On Sampling Designs with Ordered Conditional Inclusion Probabilities. Scandinavian Journal of Statistics, 40, 724-733.

3. Sinha, B.K. (1973) On Sampling Schemes to Realize Preassigned Sets of Inclusion Probabilities of First Two Orders. Calcutta Statistical Association Bulletin, 22, 89-110.

4. Tillé, Y. (2006) Sampling Algorithms. Springer, New York.

5. Brewer, K.R.W. and Hanif, M. (1983) Sampling with Unequal Probabilities. Lecture Notes in Statistics, No. 15, Springer-Verlag, New York.

6. Wywiał, J. (2000) On Precision of Horvitz-Thompson Strategies. Statistics in Transition, 4, 779-798.

7. Bondesson, L., Traat, I. and Lundqvist, A. (2006) Pareto Sampling versus Conditional Poisson and Sampford Sampling. Scandinavian Journal of Statistics, 33, 699-720.

8. Bondesson, L. (2012) On Sampling with Prescribed Second-Order Inclusion Probabilities. Scandinavian Journal of Statistics, 39, 813-829.

Chapter 5

A Probabilistic Method of Characterizing Transit Times for Quantum Particles in Non-Stationary States

Hae-Won Kim[1], Karl Sohlberg[2*]

[1]Department of Chemistry, Penn State Abington, Abington, PA, USA

[2]Department of Chemistry, Drexel University, Philadelphia, PA, USA

ABSTRACT

We present a probabilistic approach to characterizing the transit time for a quantum particle to flow between two spatially localized states. The time dependence is investigated by initializing the particle in one spatially localized "orbital" and following the time development of the corresponding non-stationary wavefunction of the time-independent Hamiltonian as the particle travels to a second orbital. We show how to calculate the probability that the particle, initially localized in one orbital, has reached a second orbital after a given elapsed time. To do so, discrete evaluations of the time-dependence of orbital occupancy, taken using a fixed time increment, are subjected to conditional probability analysis with the additional restriction of minimum flow rate. This approach yields transit-time probabilities that converge as the time increment used is decreased. The method is demonstrated on cases of two-state oscillations and shown to produce physically realistic results.

KEYWORDS

Conditional Probability; Non-Stationary Quantum Particle; Transit Time

1. INTRODUCTION

There are numerous situations where one might be interested in how quickly a quantum particle moves from one place to another. For example, electron transit times may influence such properties as binary switching speeds and alternating current conductance in molecular electronic devices. As another example, when a chemical reaction intermediate is tautomeric, the time required for the intermediate, prepared in one tautomer, to evolve into the other tautomer, may govern the overall reaction rate. In many cases this exchange of tautomers is essentially a proton transfer.

Investigating theoretically the time dependence of a quantum system begins with solving the time dependent Schrödinger equation (TDSE). Solving the TDSE yields the time dependent wavefunction $\psi(t)$ and the corresponding probability density $\rho(t)=|\psi(t)|^2$. Consider the movement of a quantum particle between two spatial regions denoted A and B: From the time dependent probability density one may calculate the probability of the particle being in region A at any given time. One may also calculate the probability of the particle being in region B at any given time, but the time dependent wavefunction does not directly yield the time it takes for the particle to travel from A to B. In order to attain quantitative knowledge of the time-dependence of the flow of a quantum particle, it is essential to develop methods to extract this information from the time-dependent wavefunction $\psi(t)$ and/or the corresponding probability density $\rho(t)$. Owing to the non-commutativity of the quantum position and momentum operators, the time required for a quantum particle to get from one spatial location to another must be defined differently from the analogous quantity for a classical particle. Here we present an approach that expresses transit time in terms of probabilistic confidences.

To date, the great majority of theoretical treatments of quantum particle flow in molecular complexes have been of the time independent variety. Notable exceptions are Baer's treatment of anti-coherence in molecular electronics [1] and Pacheco and Iyengar's demonstration of how to extract the transmission probability from the time-dependence of electron density in donor-bridgeacceptor systems [2]. In related work, Guo has applied non-perturbative time-dependent quantum dynamics to study the phenomenon of superexchange in electron transfer [3], superexchange today being of central relevance in molecular transport junctions [4,5]. Coalson has investigated the accuracy of time-dependent perturbation theory methods to treat nuclear-electronic coupling in long-range electron transfer by benchmarking results for a model Hamiltonian to its exactly solvable limit [6]. Such long-range electron transfer is relevant in a wide variety of donor-bridge-acceptor systems [7]. These studies give insight into the time dependence of electron transfer, but none recover the transit time.

A special case from the class of problems encompassed by the question, "How fast does a quantum particle move from A to B?" is the question of tunneling time through a potential barrier. Tunneling time has been a matter of considerable discussion almost since the inception of quantum mechanics. Valuable reviews have been presented by Hauge and Støvneng [8] and more recently by Winful [9]. The latter review is especially valuable for its appendix, which addresses conceptual questions about tunneling. One issue of sustained interest, apparent superluminal transport across the tunneling barrier, now appears largely resolved [10], as it has been demonstrated that causality is not violated [11,12].

Although we consider a more general problem, the transit of a quantum particle from one spatial location to another, two "tunneling time" papers are more than tangentially relevant to the present work: First, Dumont and Marchioro [13] have given a clear explanation why the question, "How much time does a tunneling particle spend under the barrier?" is ill posed. The difficulty with this question is that its answer would require the simultaneous evaluation of observables corresponding to non-commutative operators. Dumont and Marchioro argue, as we alluded to above, that the question must be posed in a probabilistic manner. Second: Baskin and Sokolovskii [14] addressed tunneling time by generalizing the classical concept of transit time to the quantum mechanical case. We too have found it useful to reference a classical case, but rather than generalizing from a deterministic classical system, we draw analogy to a probabilistic classical system.

In this paper, we present a new way to characterize transit times by analyzing the time-dependent probability density. We then demonstrate the method on the two different systems that display two state oscillations. The movement of the quantum particle is investigated by following the time development of a localized wavefunction, which is expanded in a basis of eigenfunctions of the time independent Hamiltonian using standard methods. From the time development of the wavefunction, we show how to calculate the probability that the particle, initially localized in one orbital, has reached a second orbital after a given elapsed time; or conversely, how to calculate the time that must elapse for the particle to reach the second orbital with a given probability. Collectively, we refer to these questions as the "electron orbital-occupancy problem". A related question, "How fast can a quantum state change with time?" (under the control of a time-dependent Hamiltonian) is the subject of a work with that title by Pfeifer [15].

2. TWO STATE OSCILLATIONS

The time dependence of two-state oscillations is a topic from introductory quantum mechanics. We discuss such oscillations here principally to define the time-dependent probability function for a specific case to be analyzed here, but also to introduce the notation that will be used in the balance of the paper.

Consider an electron that oscillates between two spatially localized states (These localized states could represent the source and drain of a two-terminal molecular electronic device under no applied potential, or the donor and acceptor in a DBA system, but for purposes of this study we consider two generic spatially localized noneigenstates). We want to find out how long it takes for an electron that is initially prepared in one such state to reach a second state. Such a non-stationary state might be prepared with shaped laser pulses [16] or a sudden change in potential [17]. Here we will not concern ourselves with how the spatially localized state is prepared, but rather with how to analyze its time-evolution. To begin the problem we first solve the time-dependent Schrödinger equation (TDSE) by expanding the spatial wave function in a linear combination of eigenfunctions of the time-independent Schrodinger equation (TISE) according to canonical procedure.

Consider a time-independent Hamiltonian $\hat{H}$ for an electron in which there are two eigenstates φ_i that obey the relations:

$$\hat{H}\varphi_1 = E_1\varphi_1,\ \ \hat{H}\varphi_2 = E_2\varphi_2 \quad . (1)$$

The TDSE for this system is:

$$i\hbar\partial\Psi/\partial t = \hat{H}\Psi \quad (2)$$

Given that $\hat{H}$ is time-independent, if we let: $\Psi = T(t)\Phi(r)$, then the system separates into time dependent and time independent parts with the solution of the time-dependent part being, $T(t) = \exp(-iEt/\hbar)$. Expanding the spatial wavefunction $\Phi(r)$ in a linear combination of eigenfunctions of $\hat{H}$, it follows that the general form of the solution to the TDSE is,

$$\Psi = T(t)\Phi(r) = \sum_k c_k \exp(-iE_k t/\hbar)\varphi_k , \quad (3)$$

where the c_k are the expansion coefficients at $t=0$. Since the basis functions are eigenstates, the probability of finding the system in a particular eigenstate at time t is, $\rho_k(t) = c_k \cdot c_k^*$, which is, in fact, independent of time. For our purposes, however, it is more useful to consider an expansion in non-eigenstate functions. For example, suppose that we describe a localized electron in a molecule as an expansion in atomic orbitals (AOs). We can follow the flow of charge by observing how the AO expansion coefficients vary as the time-dependent wavefunction evolves in time. Ultimately, we want to address the question; how long does it take for an electron that is initially localized in one AO to reach a second selected AO?

Suppose that the two eigenstates (φ_1 and φ_2) of the two-state system are expanded in a basis of two AOs; χ_1 and χ_2. In this basis, the TISE in secular form is,

$$\det\begin{vmatrix} \alpha - E & \beta - Es \\ \beta - Es & \alpha - E \end{vmatrix} = 0 \quad (4)$$

where $\alpha = \langle \chi_i | \hat{H} | \chi_i \rangle$, $\beta = \langle \chi_i | \hat{H} | \chi_j \rangle$, and $s = \langle \chi_i | \chi_j \rangle$. The solutions are,

$$E_1 = (\alpha + \beta)/(1+s); \text{ for which } \phi_1 = \chi_1 + \chi_2 \tag{5a}$$

$$E_2 = (\alpha - \beta)/(1-s); \text{ for which } \phi_2 = \chi_1 - \chi_2 \tag{5b}$$

If we use the approximation, $s = \delta_{ij}$, the roots are,

$$E = \alpha \pm \beta. \tag{6}$$

Both eigenvalues are displaced from the site energy α by an amount dependent on the coupling matrix element β. Since α just sets the zero of potential, we can take the eigenenergies to be $E = \pm\beta$.

Suppose that we initially localize our electron by placing it exclusively in the AO χ_1. At $t = 0$ the spatial wavefunction, written as an expansion in AOs, is,

$$\Phi(r, t=0) = \chi_1 = 1\chi_1 + 0\chi_2. \tag{7}$$

To write the time-dependent wavefunction, we must express this in the basis of eigenfunctions of $\hat{H}$. In other words, we must find the c_k in Equation (3). This is a standard change from an AO basis to a molecular orbital (MO) basis. From the definitions of ϕ_k in Equations (5a) and (5b), it follows that $c_1 = 0.5$ and $c_2 = 0.5$.

We can now write the time dependent wavefunction:

$$\begin{aligned}\Psi &= T(t)\Phi(r) \\ &= 0.5\exp(i\beta t/\hbar)\varphi_1 + 0.5\exp(-i\beta t/\hbar)\varphi_2\end{aligned} \tag{8}$$

To follow the time evolution of the wavefunction in space, we need to transform from the MO basis back to the AO basis. Using the Euler relations, we obtain the time dependent AO-occupancy probabilities as:

$$\rho(\chi_1) = \cos^2(\beta t/\hbar) \tag{9a}$$

$$\rho(\chi_2) = \sin^2(\beta t/\hbar) \tag{9b}$$

As expected, $\rho(\chi_1)+\rho(\chi_2)=1$. Note that the frequency of oscillation between the two functions is $\beta/\hbar$. Note also that the oscillation frequency depends on the energy difference between the states, the greater the energy difference, the greater the frequency.

In oscillatory systems the period of oscillation (τ) bears a reciprocal relationship to the frequency $(\tau = 2\pi/\omega)$ where (ω) is the angular frequency. As shown in the example above, a bound quantum system in a non-stationary state is characterized by such a frequency, (or frequencies in a system with more than two levels). The related period(s), however, is(are) a physically unsatisfactory metric of the lifetime of the associated spatially localized state. In Section 5 we demonstrate that this quantum exchange frequency suggests a rapid exchange between tautomers in cases when one would likely have to wait an aeon for one tautomer to evolve into another. The present paper addresses this dilemma.

Solution of the TDSE as outlined in this section yields the time-dependence of occupancy for each spatially localized orbital (AO). This standard procedure is easily generalized to an arbitrary number of AO basis functions. It is a standard result in quantum mechanics and answers the question, “what is the probability that a selected AO is occupied at a specific time t?” Herein we will demonstrate how to analyze the time dependent probability function to address our central question: Given that the electron is initially localized on one atomic orbital, how much time must elapse for the electron to reach the second orbital with a given probability? In the next section we look at an analogous classical system in order to develop the methods to answer this question. The system we look at is a "professor office-occupancy problem".

3. PROFESSOR OFFICE-OCCUPANCY PROBLEM

To develop the methodology, we first consider a classical problem analogous to the "electron orbital-occupancy problem": We ask the question; given a function that describes the probability of a professor being in her office as a function of time, what is the probability that the professor has occupied her office at least once during some elapsed time period? We term this the "professor office-occupancy problem". Exploration of this classical problem allows us to identify the variables that are important in the electron orbital-occupancy problem and to determine what assumptions might be necessary to produce a physically reasonable answer. Our approach is based on evaluating the time dependent probability density $\left(|\psi(t)|^2\right)$ at a fixed time increment. We find that conditional probability analysis is required to obtain a result that converges as the time increment is decreased. The understanding gained from studying the professor office-occupancy problem is then applied to solving the electron orbital-occupancy problem. In addition to studying the time dependence of electron orbital occupancy, the method is also demonstrated for proton exchange in a double well potential.

The professor office-occupancy problem is defined as follows: We take the probability of finding a professor in her office, $\rho(t)$, to be a sinusoidal function of time with a period of 24 hours and an amplitude of 0.79; a minimum at midnight at which there is 1% probability of finding her in her office; and a maximum of 80% probability at noon. The function is shown in Figure 1 (In this analysis we assume all days, whether weekend, weekday or holiday, are equivalent. A more complicated function with 7-day or 30-day periodicity could be used, but this would only serve to obfuscate the analysis without introducing any new physics). The probability function may be written as,

$$\rho(t) = 0.405 - 0.395 \times \cos(2\pi t/24), \tag{10}$$

where t is time in hours. The area under the curve, 9.72 hours/day, is the average amount of time the professor spends in her office each day.

This system can be thought of as a two-state oscillation. The professor is either "in" her office or "not in" her office, and she oscillates between these two "states" according to the assumed probability function (10). Now let us address the question; given a specific start time, how much time must elapse for the professor to have visited her office at least once with a selected probability? One way of thinking about the question is this: If a student comes to the professor's office, how long must the student wait to be 90% confident (or 99.999% confident) that the professor will show up?

The student could show up at the professor's office and wait. Alternatively, the student could check the office periodically. As the frequency with which the student checks the office approaches infinity, the elapsed time to finding the professor must converge to the period of time the student would have to wait at the office door. Convergence of the time to achieve a given confidence level with increasing sampling frequency is therefore a necessary condition for the validity of any approach that is based on repetitive sampling of the probability function. In the next subsection we show that simple multiplication of the probabilities extracted from the probability function fails in this regard.

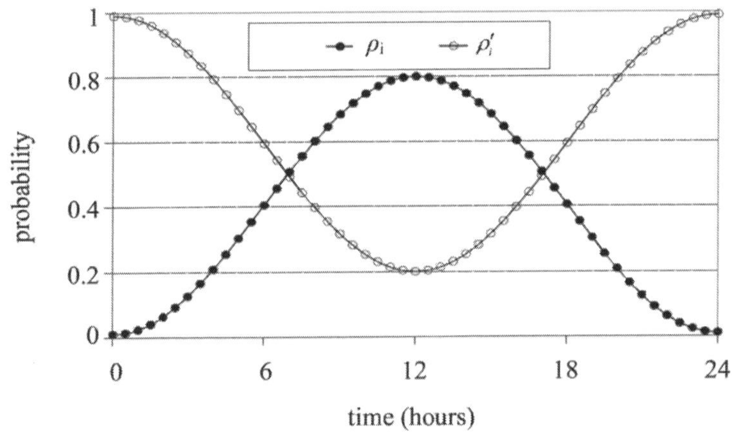

Figure 1. Probability ρ of finding a professor in her office as a function of time together with the probability ρ' that the office is not occupied.

3.1. Simple Probability Analysis

We begin our analysis of the professor office-occupancy problem by studying what happens if the student checks the professor's office periodically. The naive approach assumes probabilistic independence, in which case the probability that the office has never been found to be occupied in N samplings P_N^{not} is,

$$P_N^{not} = \prod\nolimits_{k=1,\cdots,N} \left(1 - \rho_k(t)\right), \tag{11}$$

where $\rho_k(t)$ is the probability of the office being occupied in a specific sampling event k. The probability that the professor has been found in her office is then $\left(P_N = 1 - P_N^{not}\right)$. This approach may appear to give reasonable results for widely separated sampling events $\left(\Delta t = t_k - t_{k-1} \gg 0\right)$, but obviously the assumption that $\rho(t)$ and $\rho(t - \Delta t)$ are probabilistically independent is flawed. Consequently, the method fails as $\Delta t \to 0$. This failure can be seen by considering four cases: 1) The student looks for the professor at exactly noon each day. 2) The student looks for the professor at noon and midnight each day. 3) The student looks for the professor at noon and 2 seconds after noon each day. 4) The student looks for the professor at noon and every 2 seconds thereafter until 12:00:08 pm each day.

Applying Equation (11) to the four cases described above yields the results in Table 1, which shows the day on which the probability of having found the professor exceeds 99.999%. In case 1, after N consecutive days the probability that the student has located the professor is $P_N = 1-(0.2)^N$. This expression converges to 1.0 with increasing N, a physically sensible result. In case 2, after M consecutive days the probability that the student has located the professor is $1-(0.198)^M$. This converges to 1.0, but slightly more rapidly than if the student checks only at noon, again a physically sensible result. By checking at midnight in addition to noon, the student increases the probability of finding the professor, but only by a little because the professor is very unlikely to be found in her office at midnight.

For case 3, since

$$\left(\rho'(t=12:00:00) \approx \rho'(t=12:00:02) \approx 0.2\right)$$

after N consecutive days, the probability of having located the professor will be $1-(0.2)^{2N}$. This result is completely unreasonable. The probability is converging to 1.0 more rapidly than in case 1, only because the student is making two attempts to find the professor at essentially the same time each day! Case 4 further highlights the failure of this approach. Table 1 shows that the professor is found 10 days earlier simply because the student is looking into the office every two seconds for 8 minutes each day, instead of simply checking once.

Table 1. The day the professor is found in her office with 99.999% probability, based on looking for her at different times of the day under the assumption that these samplings possess probabilistic independence.

	Formula to calculate the probability the professor has been found in her office.	Day the probability of finding the professor in her office reaches 99.999%.
Case 1	$1 - 0.2^N$	11th day
Case 2	$1 - (0.2 \times 0.99)^N$	11th day
Case 3	$1 - (0.2 \times 0.2)^N$	6th day
Case 4	$1 - (0.2 \times \cdots 0.2)^N$	2nd day

The reason for the failure of Equation (11) is that $\rho(t)$ is very strongly correlated to $\rho(t-\Delta t)$ for small Δt owing to the continuity of the time evolving probability function $\rho(t)$. In the next section we describe an approach that accounts for the correlation between sampling events to produce a predicted time to achieve a given confidence level that converges with increasing sampling frequency.

3.2. Conditional Probability

During a discrete time-step (Δt), the condition of the professor being in the office will either change or stay the same. In general, the probability that the condition changes may depend on the initial state. We can represent the evolution during a time-step with a tree as depicted in Figure 2.

In the tree diagram, a is the probability that the "state" changes given that the professor is not in her office and b is the probability that the state changes given that she is in her office ($a' = 1 - a$ and $b' = 1 - b$). We can write the probabilities of being in the office at $t + \Delta t$ as:

$$\rho_{i+1} = X + Y = \rho_i' \cdot a + \rho_i \cdot (1 - b), \tag{12a}$$

$$\rho_{i+1}' = W + Z = \rho_i' \cdot (1 - a) + \rho_i \cdot b. \tag{12b}$$

If a and b can be determined from ρ_i and ρ_{i+1} (and ρ_i' and ρ_{i+1}') it is then possible to determine the probability that the professor was never in her office. The probability is given by the probability that the office was not occupied at time $t = 0, (\rho_0')$ times the probability that the state did not change on each subsequent time-step $(a' = 1 - a)$. In other words, given that the office was not occupied at time $t = i$, if the probability that the state did not change between time $t=i$ and time $t=i+1$ is given by $a'_{i,i+1}$, then the probability that the office was never occupied is given by,

$$P^{not} = \rho'_0 \prod_{i=0,\cdots,N} a'_{i,i+1} = \rho'_0 \prod_{i=0,\cdots,N} \left(1 - a_{i,i+1}\right)$$

(13)

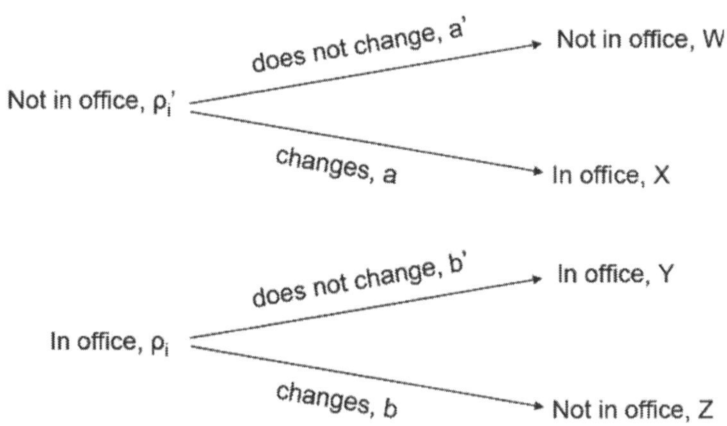

Figure 2. Conditional probability tree diagram where: ρ_i is the probability that the professor is in her office at time $t=i, \rho'_i$, is the probability that she is not in her office at time $t=i$, and $\rho'_i = 1-\rho_i$, at any time t_i.

In fact, Equations (12a) and (12b) are dependent and we can therefore only solve for a in terms of b. While a definitive relationship between a and b is not obvious, both a and b values must be confined by $0 \le (a,b) \le 1$ because they represent probabilities. By rearranging Equations (12a) and (12b), we obtain the following expressions.

$$a = (\rho_i / \rho_i')b + (\rho_i' - \rho_{i+1}') / \rho_i' \tag{14a}$$

$$b = (\rho_i' / \rho_i)a + (\rho_{i+1}' - \rho_i') / \rho_i \, . \tag{14b}$$

Using probability function Equation (10) and the relationships of (14a) and (14b), we can find the range of allowed values of a and b at each time-step.

Consider as a specific example, the set of conditions: $\rho_i = 0.6$, $\rho_i' = 0.4$, $\rho_{i+1} = 0.7$, $\rho_{i+1}' = 0.3$. It follows that for this specific set of conditions,

$$b = (4/6)a - 1/6 \tag{15}$$

We haven't identified another relationship between a and b, but we do know that $[0 \le a \le 1]$ and $[0 \le b \le 1]$. The range of allowed values of a and b for this example case is shown with the boldface line segment on the LHS graph of Figure 3.

Based on analyzing the allowed ranges for a and b, the minimum and maximum allowed values of $a(b = 0,1)$ and $b(a = 0,1)$ using two sampling timesteps, (0.5 hour and 0.1 hour) are shown in Figures 4(a) and 4(b). From the figures we see that $a_{\min}$ and $b_{\min}$ decrease as the time-step gets smaller. This is a physically sensible result: As the time step decreases, the probability that the state changes during the time step also decreases (i.e., the state is very unlikely to change during a very short time interval).

There are many mathematically allowed solutions of Equation (14) for a and b. To obtain the physically meaningful solution, a second condition on a and b is required. One obvious possibility is $a + b = 1$. Using this condition, direct application of Equation (13) produces the same unphysical result as the simple probability approach. The predicted time to achieve a given confidence level does not converge with decreasing time-step. We conclude that this obvious additional condition on a and b is not the physical one.

We note that physically, a and b represent the probabilities of the "occupied" and "unoccupied" states changing during the given time interval. While we do not know the specific values of a and b for any given time interval, it is physically reasonable that as the time step is decreased, the probability of the state changing during any one time step decreases. In the limit of decreasing time step therefore, a and b should take on their smallest allowed values. We therefore used the slowest mathematically allowed flow of probability by using $a+b=\text{minimum}$ where we denote $a=a_{\text{min}}$ and $b=b_{\text{min}}$. The case $(a+b=\text{minimum})$ represents the fewest office occupancy state changes that produce ρ_i' and ρ_{i+1}', from ρ_i and ρ_{i+1}. This is the physically meaningful additional condition on a and b.

In the range $[0<t<12]$, the probability that the office is occupied is increasing, therefore the minimum change is accomplished by setting $b=0$, which produces $a=a_{\text{min}}$. Here, if the office is occupied it remains so, if it is not, there is a small probability of a change in the state (a_{min}). Similarly, for $[12<t<24]$, the probability that the office is occupied is decreasing. The minimum change is accomplished by setting $a=0$, which produces $b=b_{\text{min}}$. In this case, if the office is empty it remains so, but if it is occupied there is a small probability of a change in the state (b_{min}).

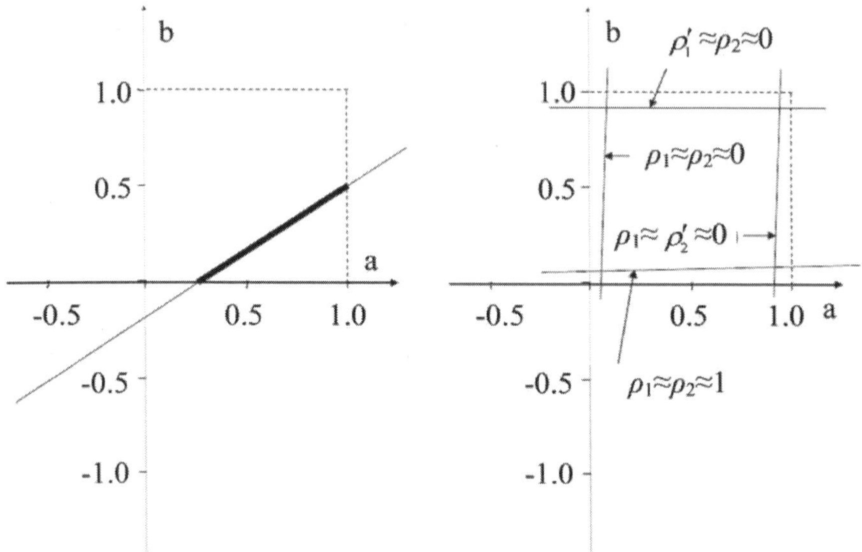

Figure 3. LHS: allowed ranges for a and b for the set of conditions: $\rho_i = 0.6$, $\rho_i' = 0.4$, $\rho_{i+1} = 0.7$, $\rho_{i+1}' = 0.3$. RHS: extrema of allowed values of a and b in general.

One might argue that the state of the system could change during the first time interval (or any subsequent time interval) regardless of how short the time increment is, in which case $a = 1$, not its minimum allowed value. This is true, however, we seek to place a confidence on the office occupancy within some elapsed time. While the rapid state change may happen, to specify a confidence that the state change has happened we must base our analysis on the slowest rate of state change that is consistent with the probability function.

Using the above analysis, we can put rigorous upper and lower bounds on the flow of probability. To obtain the upper limit, we use $a = a_{\max}$. In this case the office is occupied the first time we look. To obtain the lower limit we use $a = a_{\min}$, which produces $P^{not}(t)$ as shown in Figure 5. Note that P^{not} drops below 0.1 at 30.5 hours.

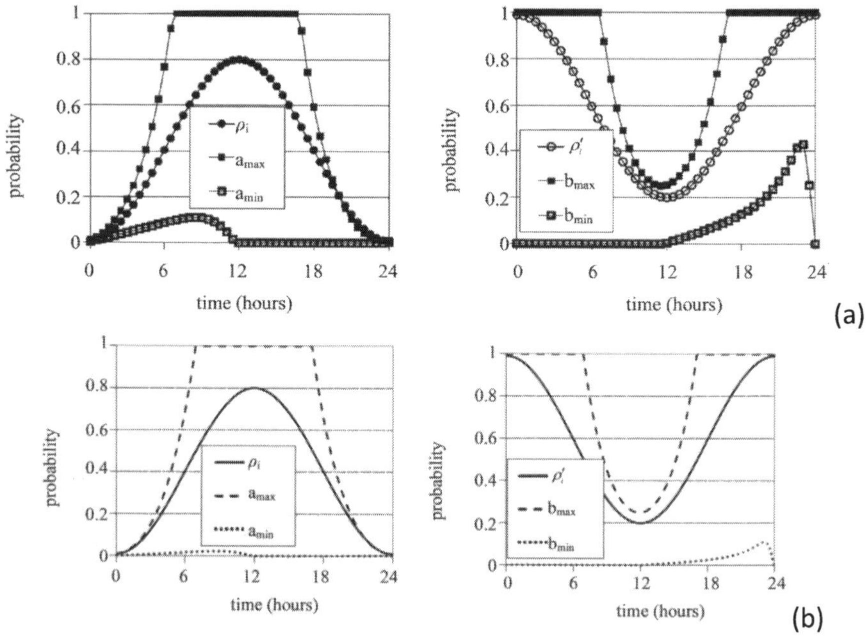

Figure 4. (a) Allowed range of a and b values for conditional probability for professor office-occupancy problem. Also plotted are the $\rho(t)$, with $a_{\max}$, $a_{\min}$ and $\rho'(t)$ with $b_{\max}$, $b_{\min}$. The plots are made for 0.5 hour time intervals. (b) Allowed range of a and b values for conditional probability for professor office-occupancy problem. Also plotted are the $\rho(t)$, with $a_{\max}$, $a_{\min}$ and $\rho'(t)$ with $b_{\max}$, $b_{\min}$. The plots are made for 0.1 hour time intervals. The spacing of data points is so small that individual points have been replaced with an interpolating curve for visual clarity. Note that $a_{\min}$ and $b_{\min}$ are smaller than in Figure 4(a) where a larger time step was used.

(Half-hour time steps were used to generate the figure). This is the elapsed time at which we can be 90% confident that the professor has been in her office at least once.

The results (collected in Table 2) show that as the time-step gets smaller, the approaches using simple probability, and conditional probability using $a+b=1$, predict that the time to 90% occupancy probability decreases to zero with increasing sampling frequency, a physically nonsensical result. Only the conditional probability analysis with the assumption of minimum flow produces a result that converges.

3.3. V alidation by Monte Carlo Simulation

3.3.1. Simulation Methods

To validate our findings, we performed Monte Carlo (MC) simulations based on the time-dependent probability function (10) for both the simple probability and conditional probability approaches. We compared the results to direct application of equations (11—simple probability analysis) and (13—conditional probability analysis). For consistency, in all cases we determined the elapsed time when the professor is 90% likely to have been in her office at least once (τ_{90}).

In the first set of MC simulations, (simple probability) the "1st visit time" is found by starting at time $t=0$ of the first day and advancing in discrete time steps Δt. At each time-step a random number is chosen between 0 and 1. If the random number falls under the probability curve $\rho(t)$ the professor is taken to be in her office and the answer is obtained. If the random number falls on or above the curve, another time-step is taken and so on until we get the 1st visit. We look at all first visits and find the time of day when 90% of the first visits have occurred. We continued to follow the professor's movement beyond the 1st visit to validate whether the simulation procedure reproduces the $\rho(t)$ profile. (This allows us to look at the office-occupancy problem without an artificial day-cycle boundary.) One or multiple consecutive in-office-states constitute a single visit. Most of the visits are in the middle of the day-cycle as expected. This approach reproduces the $\rho(t)$ profile with the professor spending on average 9.72 hours per day in the office as shown in Figure 6.

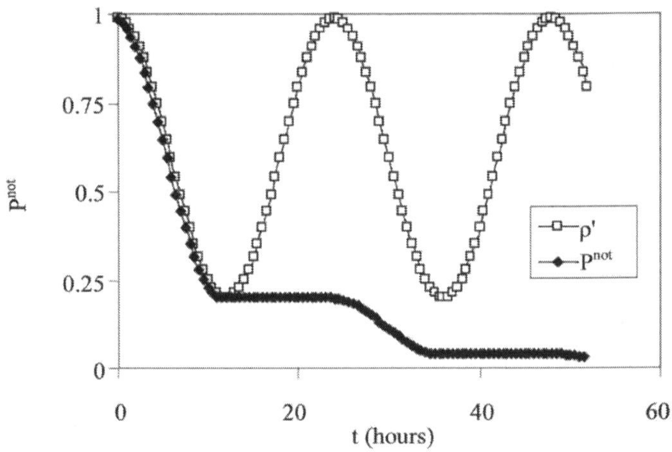

Figure 5. Time dependence of ρ' (the probability that the office is not occupied) and P^{not} (the probabilistic confidence that the office has never been occupied between $t=0$ and t) based on conditional probability analysis.

For comparison, for direct application of Equation (11), the time the professor is found in her office at least once with 90% probability is determined by setting the value of Equation (11) to 0.1 and solving for t. Here k runs from 0 to N and $t = k \cdot \Delta t$. The probability is given by the probability that the professor was not in her office at time $t = 0$, (ρ'_0) times the probability that she was not in her office at the end of each subsequent time-step (ρ'_i). The product is carried out stepwise, incrementing N by 1 until P_N^{not} drops below 0.1.

Table 2. The elapsed time (hours) at which the professor is 90% likely to have been in her office at least once (τ_{90}). The boldface result is the physically meaningful one. The other results are based on the flawed assumption that samplings of the probability function are independent events. Monte-Carlo denotes the result of a Monte-Carlo simulation. Direct denotes direct application of the noted equation.

approach →	Simple probability (Equation (11))		Conditional probability (Equation (13))		
time-step (hour) ↓	τ_{90} direct	τ_{90} Monte-Carlo	τ_{90} $a+b=1$ direct	τ_{90} a_{min}, b_{min} direct	τ_{90} a_{min}, b_{min} Monte-Carlo
60 min (1 hr)	9.00	8.00	8.00	31	30.5
30 min (1/2 hr)	6.50	6.00	6.00	31	30.7
15 min (1/4 hr)	5.00	4.75	4.75	31	30.8
10 min (1/6 hr)	4.33	4.17	4.17	31	30.9
1 min (1/60 hr)	1.68	1.65	1.68	**31.0**	30.9

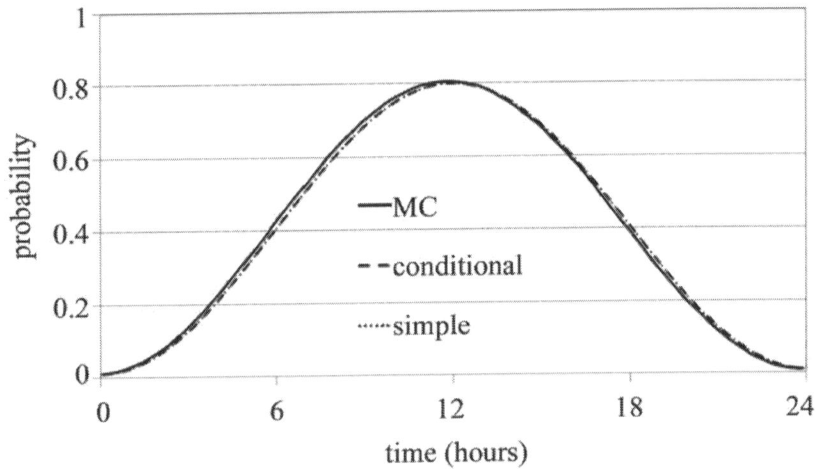

Figure 6. Comparison of $\rho(t)$ profiles based on simulations with $\rho(t)$ function of Equation (1): MC = MonteCarlo simulation, conditional = simulation with conditional probability with minimum flow, simple = simulation based on simple multiplication of probabilities. For the MonteCarlo simulations 10,000 day-cycles were used. All of the simulations reproduce the original $\rho(t)$ function and predict an average of 9.72 hours spent in office per day.

A Monte Carlo simulation based conditional probability with minimum flow is obtained in a very similar manner as the simple probability approach described two paragraphs above, but with one difference; at each timestep we compare the random number to the $a_{\min}$ in the range $0 < t < 12$ and $b_{\min}$, for the range $< 24\ 12 < t$, instead of comparing the random number to $\rho(t)$ as done in the simple probability approach.

Direct application of Equation (13) was accomplished analogously to direct application of Equation (11). This product, and all other simulations described in this manuscript, were carried out using program MATLAB [18].

3.3.2. MC Results

In Table 2 we report the results of our calculation using varying time-steps. The results show that the answer given by the simple probability approach depends on how frequently the office is checked and converges to zero as the time step is decreased, a physically nonsensical result. Simply checking the office with infinite frequency does not guarantee that the professor will arrive immediately!

As shown in Table 2, the simulation based on the condition of minimum flow produces an answer that is independent of time-step and is exactly the same as the analytic analysis shown in Figure 5, and the direct application of Equation (13). The minimum flow conditional probability approach also produces the correct $\rho(t)$ profile as shown in Figure 6 with a total office time of 9.72 hours per day-cycle.

In the approach using conditional probability with minimum flow rate, since $a = a_{\min} = 0$ from noon till midnight, no first visits are found during the final 12 hours of a day-cycle. As a result, some days have no visits at all and the answer, time 'till the first visit, is greater than 24 hours (It is important to note that the professor can arrive at her office within the first 24 hours, or 12 hours or any shorter time interval, but 90% confidence is not achieved until later). This approach achieves our requirement of obtaining a solution to the problem that converges with decreasing time-step.

There is an important feature of quantum mechanics that merits note here: In a quantum system, the timedependent wavefunction (and its corresponding probability density $\rho(t)$) evolves according to the TDSE until the state of the system is observed, whereupon the system collapses into a stationary state. In practice therefore, the process that we draw analogy to, systematically and repeatedly peeking into a professor's office to ascertain its occupancy, would function for that classical case, but not for a quantum mechanical system. In a quantum system, the first such "measurement" would collapse the wavefunction into a stationary state. It is important to draw a distinction, however, between carrying out a measurement of the system and evaluating a probability density function. To analyze the mathematical properties of $\rho(t)$ we may evaluate it at numerous values of t without altering the function, just as integrating the absolute square of a 1D spatial wavefunction $|\psi(x)|^2$ over the interval $[-\infty \leq x \leq \infty]$ to establish that it is properly normalized does not collapse the system into a single-valued position state, which would be represented by a delta function. For this reason, we may use the analogy to the classical office occupancy case to inform us of how to analyze the time evolving probability function. In the ensuing sections we apply the minimum flow conditional probability analysis to two quantum mechanical systems.

4. ELECTRON ORBITAL-OCCUPANCY

We have applied the methods developed above to the orbital-occupancy problem where $\rho(t) = \sin^2(\beta t/\hbar)$ gives the time dependence of electron occupancy for orbital χ_2 (See Equation (9b)). The oscillation frequency is $\beta/\hbar$. In our calculations we took the values of β and $\hbar$ to be 1. The cycle length is therefore π (The function repeats every time t advances by π). The results of our studies for the electron orbital occupancy problem are collected in Table 3, which shows the elapsed time (fraction of a cycle) at which the electron is 90% likely to have occupied the second orbital at least once. Note that just as in the professor office-occupancy problem, the simple probability approach produces the physically unreasonable prediction that as the sampling rate increases, the time to 90% occupancy probability converges to zero. The simple probability and $a+b=1$ Monte Carlo simulations produce the same unphysical result because they are based on the same flawed assumption of uncorrelated probabilities. The approach of using conditional probability with minimum flow rate converges to 1.25, indicating that at about 125/314th of an oscillation cycle, the electron has reached the second "acceptor" orbital with 90% probability.

While the results presented here employ a basis of two atomic orbitals, the extension to a larger basis is straightforward. Numerical integration of the TDSE for a non-stationary state representing an initially localized electron yields a discrete representation of the time dependent state vector. Upon transformation from the basis of eigenstates of the TISE to the AO basis, the squares of the AO expansion coefficients are the time dependent occupancy probabilities. Application of the conditional probability analysis described above to a selected orbital will give the elapsed time to expected occupancy of that orbital.

While our example analyses the time dependent probability function arising from a non-stationary wavefunction evolving under the influence of a time-independent Hamiltonian, it can, in principle, be applied more generally. The time-evolving probability need not be based on a time-independent Hamiltonian. In the case of a timedependent Hamiltonian, the methods of finding the time-dependent wavefunction would be different, but our method of analyzing the probability function would still apply.

5. PROTON TRANSFER ACROSS AN ASYMMETRIC DOUBLE WELL

As a second application of our conditional probability analysis we consider the exchange of a proton across an asymmetric double-well potential. Here we employ the quantum mechanical model of Hameka and de la Vega [19]. The Hamiltonian is taken to be time-independent and of the form,

$$\hat{H} = H^0 + H' . \tag{16}$$

Here H^0 describes a symmetric double-well potential having the lowest two eigenenergies $E^0 \pm \delta$, where δ represents half the splitting of these lowest two eigenenergies. The fundamental oscillation frequency is $\omega_0 = 2\delta/\hbar$. The perturbation term, $H' = \Delta \frac{x}{d}$, introduces the asymmetry to the potential. Here Δ is the energy difference between the classical minima of the asymmetric double well, d is their spatial separation and x is the displacement coordinate. Based on this model, Hameka and de la Vega [19] define a general asymmetry parameter $\rho = \Delta/2\delta$, and show that the oscillation frequency (ω) as a function of the asymmetry parameter is,

$$\omega = \omega_0 \sqrt{1+\rho^2} \quad (17)$$

A traditional measure of lifetime in quantum systems is, $\tau = 2\pi/\omega$. By this measure, the lifetime decreases with increasing asymmetry (increasing $\rho > 1$). We have posed the question; given that the deeper well A is initially occupied, how long must one wait to be 90% confident that the proton has sampled shallower well B? (As before, we denote this period τ_{90}). A plot of τ_{90} versus the asymmetry parameter is given in Figure 7, where it is compared to the traditional measure of lifetime $\tau = 2\pi/\omega$.

Table 3. The time (a.u.) at which the electron is 90% likely to have occupied the second orbital at least once. The boldface result is the physically meaningful one. The other results are based on the flawed assumption that samplings of the probability function are independent events. The oscillation frequency is $\beta/\hbar$, where $\beta=1$ and $\hbar=1$. The cycle length is π.

approach →	Simple probability			Conditional probability	
time-step (a.u.) ↓	τ_{90} direct	τ_{90} by Monte-Carlo Simulation (10,000 cycles sampled)	τ_{90} $a+b=1$ direct	τ_{90} a_{min}, b_{min} direct	τ_{90} by Monte-Carlo Simulation a_{min}, b_{min} (10,000 cycles sampled)
3.14×10^{-1} (1/10 π)	1.57	1.26	1.26	**1.26**	**1.10**
3.14×10^{-2} (1/100 π)	0.63	0.60	0.60	**1.26**	**1.24**
3.14×10^{-3} (1/1000 π)	0.28	0.28	0.28	**1.25**	**1.25**

Values of the asymmetry parameter corresponding to some of the proton transfer systems considered by Hameka and de la Vega are marked on the plot for reference. Note that even though the oscillation frequency increases with increasing asymmetry, the duration one must wait for the proton to migrate from one well to the other also increases. This is because, even though the frequency of oscillation increases, the probability of any one well being occupied does not oscillate between 0 and 1, but within a narrower range. In the limit of great asymmetry, probability is pooled almost exclusively in a single well and one must wait essentially forever for the proton to migrate, even though the oscillation frequency is very high. This is consistent with the experimentally known lack of tautomerism in 2-methylnaphthazarin [19], for which the asymmetry parameter is 9.3 × 105.

6. CONCLUSIONS

A probabilistic method is presented to compute a transit time for the transport of a quantum particle from one spatially localized state to another. This approach is based on application of conditional probability to discrete sampling of the time dependent probability function (the absolute square of the time-dependent wavefunction), with the additional condition of minimum flow rate. The method is developed by analyzing a probabilistic classical system and is subsequently applied to two cases of quantum two-state oscillations: electron orbital-occupancy and proton transfer. In this approach, the quantum particle is initially localizing in one spatially localized state (orbital) and the time development of the corresponding non-stationary wavefunction of the time-independent Hamiltonian is followed as the particle travels to a second spatially localized state. There is no definitive time at which the particle can be said to have arrived at the second orbital. Instead, we compute the elapsed time to achieve a set probabilistic confidence level (e.g. 90%, 99.999%) that the particle has reached the second state. Unlike approaches based on the (flawed) assumption that the probability that a given orbital is occupied at time t is independent of the probability that the same orbital is occupied at an earlier time t-Δt, the approach yields an answer that converges with decreasing sampling timestep. Application of the method to asymmetric proton transfer yields results that are both consistent with known experimental evidence of tautomerism and more physiccally relevant than the traditional $\tau = 2\pi/\omega$ measure of life-time. The method of calculation developed here gives a new time-dependent way to analyze and quantify the transport of quantum particles.

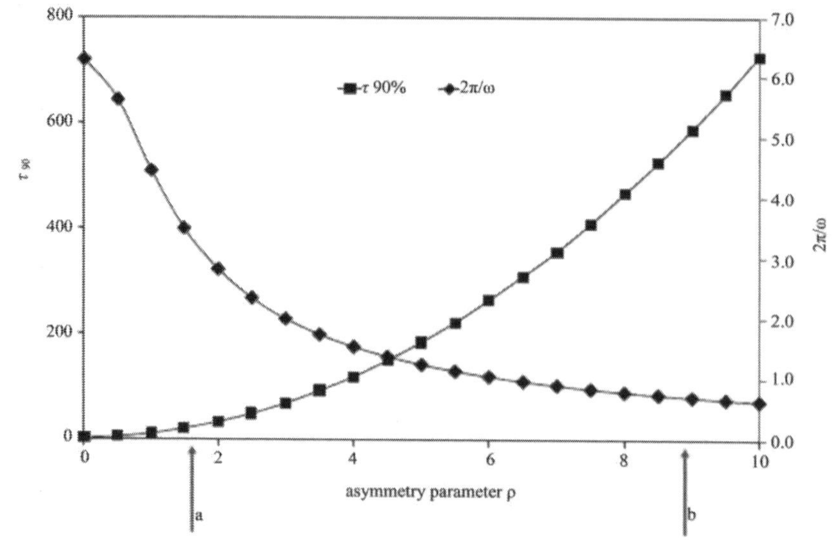

Figure 7. Dependence of τ_{90} on the asymmetry parameter (squares—left vertical axis) in comparison to the traditional measure of quantum lifetime $\tau = 2\pi/\omega$ (diamonds—right vertical axis). Atomic units are used. Arrow a marks the asymmetry parameter of 9-hydroxyphenalen-1-one and b marks the parameter for α-methyl-β-hydroxyacrolein. For comparison, the asymmetry parameter for 2-methylnaphthazarin is 9.3 × 105.

ACKNOWLEDGEMENTS

KS thanks C. Rosenthal for numerous valuable discussions as well as and K. Shuford, M. Wander, L. S. Penn and Y. Chen for critical reading of the manuscript. This work was supported by the National Science Foundation grant CHE0449595, including a ROA supplement to support Prof. Kim.

REFERENCES

1. R. Baer and D. Neuhauser, Chemical Physics, Vol. 281, 2002, pp. 353-362.

2. B. Pacheco and S. S. Iyengar, Journal of Chemical Physics, Vol. 133, 2010, Article ID: 044105. doi:10.1063/1.3463798

3. H. Guo, L. Liu and K.-Q. Lao, Chemical Physics Letters, Vol. 218, 1994, pp. 212-220. doi:10.1016/0009-2614(93)E1473-T

4. Joachim, Proceedings of the National Academy of Sciences, Vol. 102, 2005, pp. 8801-8808.

5. A. Nitzan, J. Jortner, J. Wilkie, A. L. Burin and M. A. Ratner, The Journal of Physical Chemistry B, Vol. 104 2000, pp. 5661-5665.

6. W. R. Cook, R. D. Coalson and D. G. Evans, The Journal of Physical Chemistry B, Vol. 113, 2009, pp. 11437- 11447.

7. J.-P. Launay, Chemical Society Reviews, Vol. 30, 2001, pp. 386-397.

8. J. A. Hauge and E. H. Støvneng, Reviews of Modern Physics, Vol. 61, 1989, pp. 917-936. doi:10.1103/RevModPhys.61.917

9. H. G. Winful, Physics Reports, Vol. 436, 2006, pp. 1-69.

10. Y. Aharonov, N. Erez and B. Reznik, Physical Review A, Vol. 65, 2002, Article ID: 052124. doi:10.1103/PhysRevA.65.052124

11. J. M. Deutch and F. E. Low, Annals of Physics, Vol. 228, 1993, pp. 184-202.

12. F. E. Low and P. F. Mende, Annals of Physics, Vol. 210, 1991, pp. 380-387.

13. R. S. Dumont and T. L. Marchioro II, Physical Review A, Vol. 47, 1993, pp. 85-97. doi:10.1103/PhysRevA.47.85

14. L. M. Baskin and D. G. Sokolovskii, Russian Physics Journal, Vol. 30, 1987, pp. 204-206.

15. P. Pfeifer, Physical Review Letters, Vol. 70, 1993, pp. 3365-3368.

16. R. J. Gordon and S. A. Rice, Annual Review of Physical Chemistry, Vol. 48, 1997, pp. 601-641.

17. Bohm, "Quantum Theory," Prentice Hall, Englewood Cliffs, 1951.

18. Mathworks, MATLAB, 1984-2010.

19. H. F. Hameka and J. R. de la Vega, Journal of the American Chemical Society, Vol. 106, 1984, pp. 7703-7705

Chapter 6

Missing data approaches for probability regression models with missing outcomes with applications

Li Qi[1] and Yanqing Sun[1]

Department of Mathematics and Statistics, The University of North Carolina at Charlotte

ABSTRACT

In this paper, we investigate several well known approaches for missing data and their relationships for the parametric probability regression model $P_\theta(Y|X)$ when outcome of interest Y is subject to missingness. We explore the relationships between the mean score method, the inverse probability weighting (IPW) method and the augmented inverse probability weighted (AIPW) method with some interesting findings. The asymptotic distributions of the IPW and AIPW estimators are derived and their efficiencies are compared. Our analysis details how efficiency may be gained from the AIPW estimator over the IPW estimator through estimation of validation probability and augmentation. We show that the AIPW estimator that is based on augmentation using the full set of observed variables is more efficient than the AIPW estimator that is based on augmentation using a subset of observed variables. The developed approaches are applied to Poisson regression model with missing outcomes based on auxiliary outcomes and a validated sample for true outcomes. We show that, by stratifying based on a set of discrete variables, the proposed statistical procedure can be formulated to analyze automated records that only contain summarized information at categorical levels. The proposed methods are applied to analyze influenza vaccine efficacy for an influenza vaccine study conducted in Temple-Belton, Texas during the 2000-2001 influenza season. **Mathematics Subject Classification** Primary 62J02; Secondary 62F12

KEYWORDS

Augmented inverse probability weighted estimator Asymptotic results Automated records Auxiliary outcome Efficiency Inverse probability weighted estimator Mean score estimation Recurrent events Vaccine efficacy Validation sample

1. INTRODUCTION

Suppose that Y is the outcome of interest and X is a covariate vector. One is often interested in the probability regression model $P_\beta(Y|X)$ that relates Y to X. In many medical and epidemiological studies, the complete observations on Y may not be available for all study subjects because of time, cost, or ethical concerns. In some situations, an easily measured but less accurate outcome named auxiliary outcome variable, A, is supplemented. The relationship between the true outcome Y and the auxiliary outcome A in the available observations can inform about the missing values of Y. Let V be a subsample of the study subjects, termed the validation sample, for which both true and auxiliary outcomes are available. Thus observations on (X,Y,A) are available for the subjects in V and only (X,A) are observed for those not in V.

It is well known that the complete-case analysis, which uses only subjects who have all variables observed, can be biased and inefficient, cf. Little and Rubin ([2002]). The issues also rise when substituting auxiliary outcome for true outcome; see Ellenberg and Hamilton ([1989]), Prentice ([1989]) and Fleming ([1992]). Inverse probability weighting (IPW) is a statistical technique developed for surveys by Horvitz and Thompson ([1952]) to calculate statistics standardized to a population different from that in which the data was collected. This approach has been generalized to many aspects of statistics under various frameworks. In particular, the IPW approach is used to account for missing data through inflating the weight for subjects who are underrepresented due to missingness. The method can potentially reduce the bias of the complete-case estimator when weighting is correctly specified. However, this approach has been shown to be inefficient in several situations, see Clayton et al. ([1998]) and Scharfstein et al. ([1999]). Robins et al. ([1994]) developed an improved augmented inverse probability weighted (AIPW) complete-case estimation procedure. The method is more efficient and possesses double robustness property. The multiple imputation described in Rubin ([1987]) has been routinely used to handle missing data. Carpenter et al. ([2006]) compared the multiple imputation with IPW and AIPW, and found AIPW as an attractive alternative in terms of double robustness and efficiency. Using the maximum likelihood estimation (MLE) coupled with the EM-algorithm ([Dempster et al. 1977]), Pepe et al. ([1994]) proposed the mean score method for the regression model $P_\theta(Y|X)$ when both X and A are discrete.

In this paper, we investigate several well known approaches for missing data and their relationships for the parametric probability regression model $P_\theta(Y|X)$ when outcome of interest Y is subject to missingness. We explore the relationships between the mean score method, IPW and AIPW with some interesting findings. Our analysis details how efficiency is gained from the AIPW estimator over the IPW estimator through estimation of validation probability and augmentation to the IPW score function. Applying the developed missing data methods, we derive the estimation procedures for Poisson regression model with missing outcomes based on auxiliary outcomes and a validated sample for true outcomes. Further, we show that the proposed statistical procedures can be formulated to analyze automated records that only contain aggregated information at categorical levels, without using observations at individual levels.

The rest of the paper is organized as follows. Section 2 introduces several missing data approaches for the probability regression model P_β $(Y|X)$, where the outcome Y may be missing. Section 3 explores the relationships among these estimators. The asymptotic distributions of the IPW and AIPW estimators are derived and their efficiencies are compared. Section 3 investigates efficiency of two AIPW estimators, one is based on the augmentation using a subset of observed variables and the other is based on the augmentation using the full set of observed variables. The procedures for Poisson regression using automated data with missing outcomes are derived in Section 4. The finite-sample performances of the estimators are studied in simulations in Section 5. The proposed method is applied to analyze influenza vaccine efficacy for an influenza vaccine study conducted in Temple-Belton, Texas during the 2000-2001 influenza season. The proofs of the main results are given in the Appendix A, while the proof of a simplified variance formula in Section 4 is placed in the Appendix B.

2. MISSING DATA APPROACHES

Consider the probability regression model P_β $(Y|X)$, where Y is the outcome of interest and X is a covariate vector. Let A be the auxiliary outcome for Y and V be the validation set such that observations on (X,Y,A) are available for the subjects in V and only (X,A) are observed for those in V $\bar{V}$ $\bar{V}$, the complement of V. In practice, the validation sample may be selected based on the characteristics of a subset, Z, of the covariates in X. We write $X=(Z,Z^c)$. For example, Z may include exposure indicator and other discrete covariates and Z^c may be the exposure time. Let (Z_i, X_i, Y_i, A_i), $i=1,\ldots,n$, be independent identically distributed (iid) copies of (Z,X,Y,A). Let $\xi_i = I(i \in V)$ be the selection indicator.

Most statistical methods for missing data require some assumptions on missingness mechanisms. The commonly used ones are missing completely at random (MCAR) and missing at random (MAR). MCAR assumes that the probability of missingness in a variable is independent of any characteristics of the subjects. MAR assumes that the probability that a variable is missing depends only on observed variables. In practice, if missingness is a result by design, it is often convenient to let the missing probability depend on the categorical variables only. There is also simplicity in statistical inference by modeling the missing probability based on the categorical variables. We introduce the following missing at random assumptions.

MAR I: ξ_i is independent of Y_i conditional on (X_i, A_i) and ξ_i is independent of Z_i^c conditional on (Z_i, A_i).

MAR II: ξ_i is independent of (Y_i, Z_i^c) conditional on (Z_i, A_i).

Since the conditional density $f(y, z^c|\xi, z, a) = f(z^c|\xi, z, a) f(y|z^c, \xi, z, a) = f(z^c|z, a)$ $f(y|z^c, z, a) = f(y, z^c|z, a)$, MAR I implies MAR II. It is also easy to show that MAR II implies MAR.

Let $\hat{\pi}_i$ be the estimator of the conditional probability $\pi_i = P(\xi_i = 1|X_i, A_i)$, and $\hat{\pi}_i^z$ the estimator of $\pi_i^z = P(\xi_i = 1|Z_i, A_i)$. Let $S_\beta(Y|X)$ denote the partial derivatives of $\log P_\beta(Y|X)$ with respect to β. Let $\hat{E}\{S_\beta(Y|X_i)|X_i, A_i\}$ be the estimator of the conditional expectation $E\{S_\beta$

$(Y|X_i)|X_i, A_i\}$, and $\hat{E}\{S_\beta(Y|X_i)|Z_i, A_i\}$ the estimator of $E\{S_\beta(Y|X_i)|Z_i, A_i\}$. We investigate several estimators of β based on the following estimating equations with different choices of W_i:

$$\sum_{i=1}^{n} W_i = 0, \quad (1)$$

where W_i takes one of the following forms:

$$W_i^{I1} = \frac{\xi_i}{\hat{\pi}_i^z} S_\beta(Y_i|X_i) \quad (2)$$

$$W_i^{E1} = \xi_i S_\beta(Y_i|X_i) + (1 - \xi_i)\hat{E}\left\{S_\beta(Y|X_i)|Z_i, A_i\right\} \quad (3)$$

$$W_i^{E1} = \xi_i S_\beta(Y_i|X_i) + (1 - \xi_i)\hat{E}\left\{S_\beta(Y|X_i)|Z_i, A_i\right\} \quad (4)$$

$$W_i^{A1} = \frac{\xi_i}{\hat{\pi}_i^z} S_\beta(Y_i|X_i) + \left(1 - \frac{\xi_i}{\hat{\pi}_i^z}\right)\hat{E}\left\{S_\beta(Y|X_i)|Z_i, A_i\right\} \quad (5)$$

$$W_i^{I2} = \frac{\xi_i}{\hat{\pi}_i} S_\beta(Y_i|X_i) \quad (6)$$

$$W_i^{E2} = \xi_i S_\beta(Y_i|X_i) + (1 - \xi_i)\hat{E}\left\{S_\beta(Y|X_i)|X_i, A_i\right\} \quad (7)$$

$$W_i^{A2} = \frac{\xi_i}{\hat{\pi}_i^z} S_\beta(Y_i|X_i) + \left(1 - \frac{\xi_i}{\hat{\pi}_i^z}\right)\hat{E}\left\{S_\beta(Y|X_i)|X_i, A_i\right\}. \quad (8)$$

The estimator $\hat{\beta}_{I1}$ obtained by using W_i^{I1} is an IPW estimator where a subject's validation probability π_i^z depends only on the category defined by (Z_i, A_i). Because $E\{(\pi_i^z)^{-1}\xi_i S_\beta(Y_i|X_i)\} = E\{S_\beta(Y_i|X_i)\} = 0$, the estimator $\hat{\beta}_{I1}$ is approximately unbiased. The estimator $\hat{\beta}_{I2}$ obtained by using W_i^{I2} is also an IPW estimator but with the validation probability π_i depending on the category defined by (Z_i, A_i) and the additional covariate Z_i^c.

The estimator $\hat{\beta}_{E1}$ obtained by using W_i^{E1} is the mean score estimator where the scores $S_\beta(Y_i|X_i)$ for those with missing outcomes are replaced by the estimated conditional expectations given (Z_i, A_i). The estimator $\hat{\beta}_{E2}$ obtained by using W_i^{E2} is the mean score estimator where the scores $S_\beta(Y_i|X_i)$ for those with missing outcomes are replaced by the estimated conditional expectations given (X_i, A_i). The estimator $\hat{\beta}_{E2}$ is the mean score estimator in Pepe et al. ([1994]). The mean score estimator is the MLE estimator employing the EM-algorithm ([Dempster et al. 1977]) under the assumption that the auxiliary outcome is noninformative in the sense that the probability model $P_\vartheta(A|Y,X)$ is unrelated to β.

The estimator $\hat{\beta}_{A1}$ obtained using W_i^{A1} is the AIPW estimator augmented with the estimated conditional expectation $\hat{E}\{S_\beta(Y|X_i)|Z_i, A_i\}$. The estimator $\hat{\beta}_{A2}$ obtained using W_i^{A2} is the AIPW estimator augmented with the estimated conditional expectation $\hat{E}\{S_\beta(Y|X_i)|X_i, A_i\}$. The estimator $\hat{\beta}_{A3}$ is obtained using W_i^{A3}. The W_i^{A3} differs from W_i^{A2} in that the estimated validation probability is $\hat{\pi}_i$ instead of $\hat{\pi}_i^z$

Suppose that π ˆ *z i* π̂iz $\hat{\pi}_i^z$ is an asymptotically unbiased estimator of π̄ *z i* π̄iz $\bar{\pi}_i^z$ and that $\hat{E}\{S_\beta(Y|X_i)|Z_i,A_i\}$ is asymptotically unbiased of $\bar{E}\{S_\beta(Y|X_i)|Z_i,A_i\}$, where both $\bar{\pi}_i^z$ and $\bar{E}\{S_\beta(Y|X_i)|Z_i,A_i\}$ are functions of (*Z* *i* ,*A* *i*). Under MAR II, if one of the equalities, $\bar{\pi}_i^z = \pi_i^z$ and $\bar{E}[S_\beta(Y|X_i)|Z_i,A_i]\} = E[S_\beta(Y|X_i)|Z_i,A_i]\}$, holds, then

$$E\left\{(\bar{\pi}_i^z)^{-1}\xi_i S_\beta(Y_i|X_i)\right\} + E\left\{\left(1-(\pi_i^z)^{-1}\xi_i\right)\bar{E}\left[S_\beta(Y|X_i)|Z_i,A_i\right]\right\} = E\{S_\beta(Y_i|X_i)\} = 0,$$

which entails that the estimator $\hat{\beta}_{A1}$ has the double robust property in the sense that it is a consistent estimator of *β* if either $\hat{\pi}_i^z$ is a consistent estimator of π_i^z or $\hat{E}[S_\beta(Y|X_i)|Z_i,A_i]$ is a consistent estimator of Similarly, under MAR I, the estimator *β* ˆ *A*2 β̂A2 $\hat{\beta}_{A2}$ possesses the double robust property in that $\hat{\beta}_{A2}$ is a consistent estimator of *β* if either *π* ˆ *z i* π̂iz $\hat{\pi}_i^z$ is a consistent estimator of π_i^z or $\hat{E}[S_\beta(Y|X_i)|X_i,A_i]$ is a consistent estimator of $E[S_\beta(Y|X_i)|X_i,A_i]$. The estimator $\hat{\beta}_{A3}$ has similar double robust property as $\hat{\beta}_{A2}$.

3. METHOD COMPARISONS AND ASYMPTOTIC RESULTS

Let $V(X_i,A_i)$ denote the subjects in *V* with values of (*X*,*A*) equal to (X_i,A_i), $n^V(X_i,A_i)$ the number of subjects in $V(X_i,A_i)$, and $n(X_i,A_i)$ the number of subjects with values of (*X*,*A*) equal to (X_i,A_i). When *X* and *A* are discrete and their dimensionality is reasonably small, the probability

$\pi_i = P(\xi_i = 1 | X_i, A_i)$ can be estimated by $\hat{\pi}_i = n^V(X_i, A_i)/n(X_i, A_i)$. The conditional expectation $E\{S_\beta(Y|X_i)|X_i, A_i\}$ can be nonparametrically estimated based on the validation sample,

$$\hat{E}\left\{S_\beta(Y|X_i)|X_i, A_i\right\} = \sum_{j \in V(X_i, A_i)} S_\beta\left(Y_j|X_j\right)/n^V(X_i, A_i), \tag{9}$$

Under MAR I, $\hat{E}\left\{S_\beta(Y|X_i)|X_i, A_i\right\}$ is an unbiased estimator of $E\{S_\beta(Y|X_i)|X_i, A_i\}$. Now we let $V(Z_i, A_i)$ denote the subjects in V with values of (Z,A) equal to (Z_i, A_i), $n^V(Z_i, A_i)$ the number of subjects in $V(Z_i, A_i)$, and $n(Z_i, A_i)$ the number of subjects in the sample with values of (Z,A) equal to (Z_i, A_i). A nonparametric estimator $\pi_i^z = P(\xi_i = 1|Z_i, A_i)$ is given by $\hat{\pi}_i^z = n^V(Z_i, A_i)/n(Z_i, A_i)$. A nonparametric estimator of $E\{S_\beta(Y|X_i)|Z_i, A_i\}$ is given by

$$\hat{E}\left\{S_\beta(Y|X_i)|Z_i, A_i\right\} = \sum_{j \in V(Z_i, A_i)} S_\beta\left(Y_j|X_j\right)/n^V(Z_i, A_i). \tag{10}$$

Under MAR II, (Y_i, X_i) is independent of ξ_i conditional on (Z_i, A_i), then $\hat{E}\left\{S_\beta(Y|X_i)|Z_i, A_i\right\}$ is an unbiased estimator of $E\{S_\beta(Y|X_i)|Z_i, A_i\}$.

Proposition 1.

Suppose that $X=(Z,Z^c)$ and A are discrete and their dimensionality is reasonably small. Under the nonparametric estimators $\hat{\pi}_i^z = n^V(Z_i, A_i)/n(Z_i, A_i)$, $\hat{\pi}_i = n^V(X_i, A_i)/n(X_i, A_i)$, and the estimators for the conditional expectation defined in (9) and (10), the estimators $\hat{\beta}_{I1}$, $\hat{\beta}_{E1}$ and $\hat{\beta}_{A1}$ are equivalent, and the estimators $\hat{\beta}_{I2}$,

$\hat{\beta}_{E2}$, $\hat{\beta}_{A2}$ and $\hat{\beta}_{A3}$ are equivalent. However, the estimator $\hat{\beta}_{A2}$ is different from $\hat{\beta}_{A1}$ unless Z_i^c is linearly related to Z_i in which case β is not identifiable.

The results of Proposition 1 are very intriguing since research has shown that the AIPW and the mean score methods are more efficient than the IPW method. It is also intriguing that the AIPW estimators $\hat{\beta}_{A2}$ and $\hat{\beta}_{A3}$ are actually the same estimators, not affected by the validation probability. To further understand these approaches, we investigate the asymptotic properties of these methods where (X,A) are not necessarily discrete. Through the asymptotic analysis, we gain insights about what matters to the efficiency in terms of the selections of the validation sample and the augmentation function.

Suppose that $\tilde{E}\left\{S_\beta(Y|X_i)|X_i,A_i\right\}$ is a consistent parametric/nonparametric estimator of $E_a\left\{S_\beta(Y|X_i)|X_i,A_i\right\}$, where $E_a\left\{S_\beta(Y|X_i)|X_i,A_i\right\}$ is $E\left\{S_\beta(Y|X_i)|X_i,A_i\right\}$ or $E\{S_\beta(Y|X_i)|Z_i,A_i\}$. Let $\pi(X_i,A_i,\psi)$ be the parametric model for the validation probability π_i, where ψ is a q-dimensional parameter. We show in Corollary 2 that the nonparametric estimator of $\pi(X_i,A_i,\psi)$ can also be expressed in the parametric form when (X_i,A_i) are discrete. Let ψ_0 be the true value of ψ. Under MAR I, the MLE $\hat{\psi}=\left(\hat{\psi}_1,\ldots,\hat{\psi}_q\right)$ of $\psi=(\psi_1,\ldots,\psi_q)$ is obtained by maximizing the observed data likelihood,

$$\prod_{i=1}^{n}\{\pi(X_i,A_i,\psi)\}^{\xi_i}\{1-\pi(X_i,A_i,\psi)\}^{1-\xi_i}.$$

The validation probability π_i is estimated by $\tilde{\pi}_i = \pi\left(X_i, A_i, \hat{\psi}\right)$. Then by the standard likelihood based analysis, we have the approximation

$$\hat{\psi} - \psi_0 = n^{-1} \sum_{i=1}^{n} \left(I^{\psi}\right)^{-1} S_i^{\psi} + o_p\left(n^{-1/2}\right), \tag{11}$$

where $S \psi i$ Siψ S_i^{ψ} and I^{ψ} are the score vector and information matrix for ψ ˆ ψ̂ $\hat{\psi}$ defined by

$$\begin{aligned} S_i^{\psi} &= \frac{(\xi_i - \pi(X_i, A_i, \psi_0))}{\pi(X_i, A_i, \psi_0)(1 - \pi(X_i, A_i, \psi_0))} \frac{\partial \pi(X_i, A_i, \psi_0)}{\partial \psi}, \\ I^{\psi} &= E\left\{\frac{1}{\pi(X_i, A_i, \psi_0)(1 - \pi(X_i, A_i, \psi_0))} \left(\frac{\partial \pi(X_i, A_i, \psi_0)}{\partial \psi}\right)^{\otimes 2}\right\}, \end{aligned} \tag{12}$$

where $a^{\otimes 2} = a a'$.

Consider the IPW estimator $\hat{\beta}_I$ obtained by solving the estimating equation

$$U_I = \sum_{i=1}^{n} \frac{\xi_i}{\tilde{\pi}_i} S_\beta(Y_i|X_i) \tag{13}$$

and the AIPW estimator $\hat{\beta}_A$ based on solving the estimating equation

$$U_A = \sum_{i=1}^{n} \left[\frac{\xi_i}{\tilde{\pi}_i} S_\beta\left(Y_i|X_i\right) + \left(1 - \frac{\xi_i}{\tilde{\pi}_i}\right) \tilde{E}\left\{S_\beta(Y|X_i)|X_i, A_i\right\}\right]. \tag{14}$$

Theorem 1.

Assume that $P_\beta(Y|X)$ and $\pi(X,A,\psi)$have bounded third-order derivatives in a neighborhood of the true parameters and are bounded away from 0 almost surely, both $-E\{(\partial^2/\partial \beta^2)(\log P_\beta(Y|X))\}$ and I^ψ are positive definite at the true parameters. Then, under *MAR I*,

$$n^{1/2}\left(\hat{\beta}_I-\beta\right)=I^{-1}(\beta)n^{-1/2}\sum_{i=1}^{n}Q_i^I+o_p(1),$$

$$n^{1/2}\left(\hat{\beta}_A-\beta\right)=I^{-1}(\beta)n^{-1/2}\sum_{i=1}^{n}Q_i^A+o_p(1),$$

where $I(\beta)=E\{-(\partial^2/\partial \beta^2)(\log P_\beta(Y|X))\}=\mathrm{Var}(S_\beta(Y_i|X_i))$,

$$Q_i^I=\xi_i/\pi_i S_\beta(Y_i|X_i)-E\left\{\pi_i^{-2}\xi_i S_\beta(Y_i|X_i)\left(\partial\pi(X_i,A_i,\psi_0)/\partial\psi\right)'\right\}(I^\psi)^{-1}S_i^\psi$$

and $Q_i^A=\xi_i/\pi_i S_\beta(Y_i|X_i)+(1-\xi_i/\pi_i)\,E_a\left\{S_\beta(Y|X_i)|X_i,A_i\right\}$.

Both $n^{1/2}\left(\hat{\beta}_I-\beta\right)$ and $n^{1/2}\left(\hat{\beta}_A-\beta\right)$ have asymptotically normal distributions with mean zero and covariances equal to $I^{-1}(\beta)\mathrm{Var}\left(Q_i^I\right)I^{-1}(\beta)$ and $I^{-1}(\beta)\mathrm{Var}\left(Q_i^A\right)I^{-1}(\beta)$, respectively. Further,

$$Var\left(Q_i^I\right)=Var\left(Q_i^A\right)+Var(B_i+O_i) \tag{15}$$

and

$$Var\left(Q_i^A\right)=I(\beta)+Var\left(\left(1-\frac{\xi_i}{\pi_i}\right)\left\{S_\beta(Y_i|X_i)-E_a\left\{S_\beta(Y|X_i)|X_i,A_i\right\}\right\}\right), \tag{16}$$

where

$O_i = E\{\pi_i^{-2}\xi_i S_\beta(Y_i|X_i)(\partial\pi(X_i, A_i, \psi_0)/\partial\psi)'\}(I^\psi)^{-1}S_i^\psi$ *and* $B_i = (1 - \xi_i/\pi_i)$

and $E_a\{S_\beta(Y|X_i)|X_i, A_i\}$.

Suppose that the validation probability $\pi_i = P(\xi_i = 1|X_i, A_i)$ depends only on (Z_i, A_i). That is, $\pi_i = \pi_i^z = P(\xi_i = 1|Z_i, A_i)$. Suppose that $\hat{\pi}_i$ is the MLE of π_i^z under the parametric family $\psi(Z_i, A_i, \psi)$. Let $\hat{\beta}_{A1}$ be the estimator obtained by solving (14) where the augmented term, $\hat{E}\{S_\beta(Y|X_i)|X_i, A_i\}$, is a consistent parametric/nonparametric estimator of $E\{S_\beta(Y|X_i)|Z_i, A_i\}$. Let $\hat{\beta}_{A2}$ be the estimator obtained by solving (14) where $\hat{E}\{S_\beta(Y|X_i)|X_i, A_i\}$ is a consistent parametric/nonparametric estimator of $E\{S_\beta(Y|X_i)|X_i, A_i\}$. The following corollary presents the asymptotic results for two AIPW estimators of β, one that corresponds to the augmentation based on a subset, (Z,A), of observed variables and the other that corresponds to the augmentation based on the full set, (X,A), of the observed variables.

Corollary 1.

Suppose that the validation probability $\pi_i = P(\xi_i = 1|X_i, A_i)$ depends only on (Z_i, A_i). Under the conditions of Theorem 1,

$$n^{1/2}\left(\hat{\beta}_{A1} - \beta\right) \xrightarrow{\mathcal{D}} N\left(0, I^{-1}(\beta) + I^{-1}(\beta)\Sigma_{A1}(\beta)I^{-1}(\beta)\right), \quad (17)$$

and

$$n^{1/2}\left(\hat{\beta}_{A2}-\beta\right) \xrightarrow{\mathcal{D}} N\left(0, I^{-1}(\beta)+I^{-1}(\beta)\Sigma_{A2}(\beta)I^{-1}(\beta)\right),$$

(18)

where

$\Sigma_{A1}(\beta)=E\left[((1-\pi_i^z)/\pi_i^z)Var\{S_\beta(Y_i|X_i)|Z_i,A_i\}\right]$ *and* $\Sigma_{A2}(\beta)=E\left[((1-\pi_i^z)/\pi_i^z)\right.$ $\left.Var\left\{S_\beta(Y_i|X_i)|X_i,A_i\right\}\right]$. The asymptotic variance of β ˆ $A2$ $\hat{\beta}$A2 $\hat{\beta}_{A2}$ is smaller than the asymptotic variance of β ˆ $A1$ $\hat{\beta}$A1 $\hat{\beta}_{A1}$ if the covariates Z_i are a proper subset of X_i.

Suppose that (Z,A) are discrete taking values (z,a) in a set of finite number of values. If the number of parameters in ψ equals the number of values $\psi_{z,a}=P(\xi_i=1|Z_i=z,A_i=a)$ for all distinct pairs (z,a), then $\psi=\{\psi_{z,a}\}$ and $\pi(z,a,\psi)=\psi_{z,a}$. Further, $\partial\pi(z,a,\psi_0)\partial\psi$ ∂π(z,a,ψ0)∂ψ $\frac{\partial\pi(z,a,\psi_0)}{\partial\psi}$ can be viewed as a column vector with 1 in the position for $\psi_{z,a}$ and 0 elsewhere. The information matrix I^ψ defined in (12) has the expression,

$$I^\psi=\sum_{z,a}\rho(z,a)\left\{\frac{1}{\pi(z,a,\psi_0)(1-\pi(z,a,\psi_0))}\frac{\partial\pi(z,a,\psi_0)}{\partial\psi}\left(\frac{\partial\pi(z,a,\psi_0)}{\partial\psi}\right)'\right\},$$

where $\rho(z,a)=P(Z_i=z,A_i=a)$. It follows that I^ψ is a diagonal matrix and its inverse matrix is also diagonal. The MLE $\hat{\psi}_{z,a}=n^V(z,a)/n(z,a)$ is in fact the nonparametric estimator for $\psi_{z,a}$ based on the proportion of validated samples in the category specified by (z,a). The equation (11) can be expressed as

$$\hat{\psi}_{z,a}-\pi(z,a,\psi_0)=n^{-1}\frac{n^V(z,a)-n(z,a)\pi(z,a,\psi_0)}{\rho(z,a)}+o_p\left(n^{-1/2}\right),$$

for $(z,a)\in$.

By Threom 1, the possible efficiency gain of the AIPW estimator over the IPW estimator is shown through the equation (15). The AIPW estimator is more efficient unless Var(B_i+O_i)=0. In particular, from the proof of Theorem 1, we have

$$n^{-1/2}U_A = n^{-1/2}\sum_{i=1}^{n} Q_i^A + o_p(1) \tag{19}$$

$$n^{-1/2}U_I = n^{-1/2}\sum_{i=1}^{n} Q_i^A - n^{-1/2}\sum_{i=1}^{n}(B_i + O_i) + o_p(1), \tag{20}$$

where B_i and O_i are defined following (16). The following corollary presents the analysis of the term $n^{-1/2}\sum_{i=1}^{n}(B_i + O_i)$ when (Z_i, A_i) are discrete to understand how efficiency may be gained from the AIPW estimator over the IPW estimator.

Corollary 2.

Under the conditions of Theorem 1, suppose that $X=(Z,Z^c)$ and (Z,A) are discrete taking values (z,a) in a set of finite number of values. Suppose that the validation probability $\pi_i=P(\xi_i=1|X_i,A_i)$ only depends on (Z_i,A_i) and $\psi_{z,a}=P(\xi_i=1|Z_i=z,A_i=a)$ is estimated nonparametrically by $\hat{\psi}_{z,a} = n^V(z,a)/n(z,a)$. Then

$$n^{-1/2}\sum_{i=1}^{n}(O_i + B_i)$$
$$= -\sum_{z,a} n^{-1/2}\sum_{j=1}^{n}\frac{\xi_j - \pi(z,a,\psi_0)}{\pi(z,a,\psi_0)} I(Z_j = z, A_j = a)$$
$$\left[E\{S_\beta(Y_j|X_j)|Z_j = z, A_j = a\} - E\left\{S_\beta(Y_j|X_j)|Z_j = z, Z_j^c, A_j = a\right\}\right]. \tag{21}$$

By Corollary 2, (19) and (20), $\hat{\beta}_A$ is more efficient than $\hat{\beta}_I$ unless $Var\{S_\beta(Y_j|X_j)|Z_j=z,A_j=a\}=0$ for all (z,a) for which $P(Z_i=z,A_i=a)\neq 0$. If $X=Z$ and the validation probability $\pi_i=P(\xi_i=1|X_i,A_i)$ is nonparametrically estimated with the cell frequencies $\hat{\psi}_{z,a}=n^V(z,a)/n(z,a)$, then $\hat{\beta}_A$ and $\hat{\beta}_I$ are asymptotically equivalent.

Remark Consider the estimators of β obtained based on the estimating equation (1) corresponding to different choices of W_i given in (2) to (8). If (Z,A) are discrete and the validation probability $\pi_i^z=P(\xi_i=1|Z_i,A_i)$ is estimated nonparametrically by the cell frequency, then by Theorem 1 and Corollary 2, $\hat{\beta}_{A1}$ and $\hat{\beta}_{I1}$ have same asymptotic normal distributions as long as $\hat{E}[S_\beta(Y|X_i)|Z_i,A_i]$ is a consistent estimator of $E[S_\beta(Y|X_i)|Z_i,A_i]$. But $\hat{\beta}_{A2}$ is more efficient than $\hat{\beta}_{I1}$ as long as $\hat{E}[S_\beta(Y|X_i)|X_i,A_i]$ is a consistent estimator of $E[S_\beta(Y|X_i)|X_i,A_i]$ since $Var(B_i+O_i)$ is not zero by (21). These results are not affected by whether $E[S_\beta(Y|X_i)|Z_i,A_i]$ and $E[S_\beta(Y|X_i)|X_i,A_i]$ are estimated nonparametrically or based on some parametric models. In addition, by Theorem 1, Corollary 1 and 2, $\hat{\beta}_{A3}$ and $\hat{\beta}_{I2}$ have the same asymptotic normal distributions as long as $\hat{E}[S_\beta(Y|X_i)|X_i,A_i]$ is a consistent estimator of $E[S_\beta(Y|X_i)|X_i,A_i]$.

4. POISSON REGRESSION USING THE AUTOMATED DATA WITH MISSING OUTCOMES

Many medical and public health data are available only in aggregated format, where the variables of interest are aggregated counts without being available at individual levels. Many existing statistical methods for missing data require observations at individual levels. Applying the missing data methods presented in Section 3, we derive some estimation procedures for the Poisson regression model with missing outcomes based on auxiliary outcomes and a validated sample for true outcomes. Further, we show that, by stratifying based on a set of discrete variables, the proposed statistical procedure can be formulated so that it can be used to analyze automated records which do not contain observations at individual levels, only summarized information at categorical levels.

Let Y represent the number of events occurring in the time-exposure interval $[\,0,T]$ and Z the covariates. We consider the Poisson regression model,

$$P(Y = y|Z, T) = \exp\left\{-T\exp\left(\beta' Z\right)\right\}\left\{T\exp\left(\beta' Z\right)\right\}^{y}/y!, \tag{22}$$

where Z is a vector of k+1 covariates and β a vector of k+1 regression coefficients. In practice, the exact number of true events may not be available for all subjects. We may instead have the number of possible events, namely, the auxiliary events. For example, in the study of vaccine adverse events associated with childhood immunizations, the number of auxiliary events A for MAARI is collected based on ICD-9 codes through hospital records. Further diagnosis may indicate that some of these events are false events. The number of true vaccine adverse events, Y, can only be confirmed for the subjects in the validation set V. Suppose that Z is the vaccination status, 1 for the vaccinated and 0 for the unvaccinated. Then, under Poisson regression, $\exp(\beta)$ is the relative rate of event occurrence per unit time of the exposed versus unexposed. We assume that the number of automated events A can be expressed as $A=Y+W$, where W is the number of false events independent of Y conditional on (Z,T). Suppose that W follows the Poisson regression model

$$P(W = w|Z, T) = \exp\left\{-T \exp\left(\gamma' Z\right)\right\} \left\{T \exp(\gamma' Z)\right\}^{w}/w!, \tag{23}$$

where $\gamma'=(a_0,a_1,\gamma_1,\cdots,\gamma_{k-1})$.

We apply the missing data methods introduced in Section 3 on model (22). The variables (Z_i,T_i,Y_i,A_i) are observed for the validation sample V and (Z_i,T_i,A_i) are observed for the nonvalidation sample V $\bar{V}\bar{V}$. While the covariate Z can be considered as categorical, it is natural to consider the exposure time T as a continuous variable. We assume that the validation probability depends only on the stratification of (Z,A). That is, the validation sample is a stratified random sample by the categories defined by (Z,A). Of those estimators discussed in Section 2, there are only two different estimators, $\hat{\beta}_{I1}$ and $\hat{\beta}_{A2}$. We show in Section 4.3 that the proposed method can be formulated so that it can be used to analyze the automated records with missing outcomes. First we derive the explicit estimation procedures for $\hat{\beta}_{I1}$ and $\hat{\beta}_{A2}$ and their variance estimators under model (22).

4.1 Inverse probability weighting estimation

We adopt all notations introduced in Section 3. In particular, let $\pi_i^z = P(\xi_i = 1|Z_i, A_i)$ and $\hat{\pi}_i^z = n^V(Z_i, A_i)/n(Z_i, A_i)$. Let $X = (Z, T)$. Let $X=(Z,T)$ and $X_i=(Z_i,T_i)$ to be consistent with earlier notations. The score function for subject i

under model (22) is $S_\beta(Y_i|X_i) = Z_i'(Y_i - T_i\exp(\beta' Z_i))$. The estimator $\hat{\beta}_{I1}$ is obtained by solving $\sum_{i=1}^{n}(\xi_i/\hat{\pi}_i^z)S_\beta(Y_i|X_i) = 0$, where $S_\beta(Y_i|X_i) = Z_i'(Y_i - T_i\exp(\gamma' Z_i))$. By Corollary 1, $\sqrt{n}(\hat{\beta}_{I1} - \beta)$ converges in distribution to a normal distribution with mean zero and the variance $I^{-1}(\beta)\Sigma_{A1}(\beta)I^{-1}(\beta)$, where $\Sigma_{A1}(\beta) = E\left[((1-\pi_i^z)/\pi_i^z)\mathrm{Var}\left\{S_\beta(Y_i|X_i)|Z_i, A_i\right\}\right]$.

The information matrix $I(\beta) = E(Z_iZ_i'T_i\exp(\beta' Z_i)) = \sum_z P(Z_i = z)zz'\exp(\beta' z)$ $E(T_i|Z_i = z)$ can be estimated by $\hat{I}(\hat{\beta})$ which is obtained by replacing *β* with $\hat{\beta}_{I1}$, $P(Z_i = z)$ by the sample proportion of the event $\{Z_i = z\}$, and $E(T_i|Z_i = z)$ with the sample average exposure time for those with covariates $Z_i = z$. The matrix $\Sigma_{A1}(\beta)$ can be estimated by

$$\hat{\Sigma}_{A1}(\hat{\beta}) = \sum_{a,z}\hat{\rho}(a,z)\frac{1-\hat{\rho}^{v}(a,z)}{\hat{\rho}^{v}(a,z)}\widehat{\mathrm{Var}}\{S_\beta(Y|X)|A = a, Z = z\},$$

(24)

where $\hat{\rho}(a,z)$ is the estimator of $\hat{\rho}^{v}(a,z)$ is the estimator of $P\{i \in V|A_i = a, Z_i = z\}$, and $\{S_\beta(Y_i|X_i)|Z_i, A_i\}]$ which is derived in the following.

Since *A* is observed for all subjects, *W* can be determined if *Y* is known, and undetermined otherwise. The IPW estimator, $\hat{\gamma}_{I1}$, of *γ* can be estimated by solving the equation $\sum_{i=1}^{n}(\xi_i/\hat{\pi}_i^z)S_\gamma(W_i|X_i) = 0$, where $S_\gamma(W_i|X_i) = Z_i'(W_i - T_i\exp(\gamma' Z_i))$. The conditional distribution of *Y* given *A=a*, *T*, and *Z=z* is Binomial (a, p_z), where $p_z = \exp(\beta' z)/(\exp(\beta' z) + \exp(\gamma' z))$. Since this conditional distribution does not depend on *T*, the outcome *Y* and *T* are conditionally independent given (*A*,*Z*). Therefore, $\mathrm{Var}\{S_\beta(Y|X)|A,Z\} = Z$

$Z'\{\mathrm{Var}(Y|A,Z) + \exp(2\beta'Z)\mathrm{Var}(T|A,Z)\}$. The variance $\mathrm{Var}(Y|A=a,Z=z)$ can be estimated by $a\hat{p}_z(1-\hat{p}_z)$, where $\hat{p}_z = \exp(\hat{\beta}'z)/\left\{\exp(\hat{\beta}'z)+\exp(\hat{\gamma}'z)\right\}$, and $\mathrm{Var}(T|A=a,Z=z)=E(T^2|A=a,Z=z)-\{E(T|A=a,Z=z)\}^2$ can be estimated nonparametrically using the first and the second sample moments conditional on each category with $A=a$ and $Z=z$.

4.2 Augmented inverse probability weighted estimation

Under the assumption that W follows the Poisson regression model (23) and is independent of Y conditional on (Z,T), (Z,T), $E\left\{S_\beta(Y|X)|Z,T,A\right\} = AZ'\frac{\exp(\beta'Z)}{\exp(\beta'Z)+\exp(\gamma'Z)} - TZ'\exp(\beta'Z)$. Let $\hat{E}\left\{S_\beta(Y|X_i)|X_i,A_i\right\}$ be the estimator of $E\{S_\beta(Y|X_i)|X_i,A_i\}$ for a given β by substituting γ by its estimator $\hat{\gamma}_{I1}$ of Section 4.1. Then the estimator $\hat{\beta}_{A2}$ is obtained by solving

$$\sum_{i=1}^{n}\frac{\xi_i}{\hat{\pi}_i^z}S_\beta(Y_i|X_i)+\left(1-\frac{\xi_i}{\hat{\pi}_i^z}\right)\hat{E}\left\{S_\beta(Y|X_i)|X_i,A_i\right\}=0. \tag{25}$$

By Corollary 1, n √ (β ˆ A2 −β) n(βA2−β) $\sqrt{n}(\hat{\beta}_{A2}-\beta)$ converges in distribution to a normal distribution with mean zero and the variance matrix where $I^{-1}(\beta)+I^{-1}(\beta)\Sigma_{A2}(\beta)I^{-1}(\beta)$, where $\Sigma_{A2}(\beta) = E\left[((1-\pi_i^z)/\pi_i^z)\mathrm{Var}\{S_\beta(Y_i|X_i)|X_i,A_i\}\right]$. The information matrix $I(\beta)$ can be estimated by $\hat{I}(\hat{\beta})$ given in Section 4.1. The conditional variance $\mathrm{Var}\{S_\beta(Y|X)|Z=z,T,A=a\}=a\,p_z(1-p_z)z^{\otimes 2}$ can be estimated by $a\hat{p}_z(1-\hat{p}_z)$, where

$\hat{p}_z = \exp(\hat{\beta}'z)/(\exp(\hat{\beta}'z) + \exp(\hat{\gamma}'z))$. It follows that $\Sigma_{A2}(\beta)$ can be consistently estimated by

$$\hat{\Sigma}_{A2}(\beta) = \sum_{a,z} \hat{\rho}(a,z)\frac{1-\hat{\rho}^v(a,z)}{\hat{\rho}^v(a,z)} a\hat{p}_z(1-\hat{p}_z)z^{\otimes 2},$$

where $\hat{\rho}(a,z)$ is the estimator of $P\{A_i = a, Z_i = z\}$ and $\hat{\rho}^v(a,z)$ is the estimator of $P\{i \in V | A_i = a, Z_i = z\}$.

4.3 Estimation using the automated data

This section formulates the missing data estimation procedure for (22) based on the automated (summarized) information at categorical levels defined by relevant covariates of the model. In particular, we show that $\hat{\beta}_{I1}$ and $\hat{\beta}_{A2}$ and their variance estimators can be formulated using the automated data at categorical levels.

In many applications it is convenient to write $Z=(1,Z_{(1)},Z_{(2)})$ and $\beta=(b_0,b_1,\vartheta')'$, where $Z_{(1)}$ is the treatment indicator ($Z_{(1)}=1$ for the exposed group and $Z_{(1)}=0$ for the unexposed group) and $Z_{(2)}=(\eta_1,\cdots,\eta_{k-1})'$ as the other covariates, and $\vartheta=(\vartheta_1,\cdots,\vartheta_{k-1})'$. For the applications involving the automated data records, we let $\eta_1,\cdots,\eta_{k-1}$ be $k-1$ dummy variables representing k groups. Without loss of generality, we choose the k th group as the reference group, $\eta_1=1$, $\eta_2=0$, $\cdots$, $\eta_{k-1}=0$ for group 1, $\eta_1=0$, $\eta_2=1$, $\cdots$, $\eta_{k-1}=0$ for group 2, so on and $\eta_1=0$, $\eta_2=0$, $\cdots$, $\eta_{k-1}=0$ for group k. Thus each value of Z denotes a category which can be represented by (l,m) for $l=0,1$ and $m=1,\cdots,k$. This correspondence is denoted by $Z\simeq(l,m)$ for convenience. For $l=0,1$ and $m=1,\cdots,k-1$, category (l,m) is defined by Z with $Z_{(1)}=l$, $\eta_m=1$ and $\eta_j=0$ for $j\neq m$, $j=1,\ldots,k$, and category (l,k) is defined by $Z_{(1)}=l$ and $\eta_j=0$ for $j=1,\ldots,k-1$. Under model (22), the expected number of events for a subject in category (l,m) with the time-exposure interval $[0,T]$ is $T\exp(b_{lm})$, for $l=0,1$ and $m=1,\cdots,k$, where $b_{1k}=b_0+b_1$,

$b_{0k}=b_0$, $b_{1m}=b_0+b_1+\vartheta_m$ and $b_{0m}=b_0+\vartheta_m$ for $1\leq m\leq k-1$. The parameter b_1 represents the log-relative rate of the exposed versus the unexposed adjusted for other factors.

The following notations are used to show that the estimators of β and their variance estimators can be calculated using the automated information at the categorical levels. Let $V(a,l,m)$ denote the set of subjects in V with ($A=a$, $Z\simeq(l,m)$), $V(l,m)$ for the set of subjects in V with ($Z\simeq(l,m)$), n_{alm} for the number of subjects with ($A=a$, $Z\simeq(l,m)$), n^{v}_{alm} nalmv n^{v}_{alm} for the number of subjects in $V(a,l,m)$, n^{v}_{lm} nlmv n^{v}_{lm} for the number of subjects in $V(l,m)$, $\lambda_{alm}=n_{alm}/n^{v}_{alm}$ λalm=nalm/nalmv $\lambda_{alm}=n_{alm}/n^{v}_{alm}$, y_{alm} for the number of events for subjects in $V(a,l,m)$, y_{lm} for the number of events for subjects in $V(l,m)$, t_{alm} for the total exposure time for subjects with ($A=a$, $Z\simeq(l,m)$), $t_{2,alm}$ for the total squared exposure time for subjects with ($A=a$, $Z\simeq(l,m)$), t_{lm} for the total exposure time for subjects with $Z\simeq(l,m)$, α_{lm} for the number of automated events for subjects with $Z\simeq(l,m)$.

Estimation with $\hat{\beta}_{I1}$ using the automated data. The validation probability π^{z}_{i} πiz π^{z}_{i} can be estimated by $1/\lambda_{alm}$ when $A_i=a$, $Z_i\simeq(l,m)$.

It can be shown that the estimating equation for $\hat{\beta}_{I1}$ is equivalent to the following nonlinear equations for $\{b_{lm}$, for $l=0,1$, $m=1,\cdots,k\}$,

$$\sum_{m=1}^{k}\left(\sum_{a\in A} y_{alm}\lambda_{alm} - e^{b_{lm}}\sum_{a\in A} t_{alm}\lambda_{alm}\right) = 0,$$

$$\sum_{l=0,1}\left(\sum_{a\in A} y_{alm}\lambda_{alm} - e^{b_{lm}}\sum_{a\in A} t_{alm}\lambda_{alm}\right) = 0,$$

for l=0,1 and m=1,…,k−1. When k>1, the equations have no explicit solutions.

In the following, we show that the asymptotic variance of $\hat{\beta}_{I1}$ can be consistently estimated by only using the automated information at categorical levels. The information matrix is a (k+1)×(k+1) symmetric matrix given by

$$I(\beta) = E(Z_i Z_i' T_i \exp(\beta' Z_i))$$
$$= \begin{bmatrix} \sum_{l,m} q_{lm} & \sum_{m=1}^{k} q_{1m} & q_{11}+q_{01} & \cdots & q_{1r}+q_{0r} \\ \sum_{m=1}^{k} q_{1m} & \sum_{m=1}^{k} q_{1m} & q_{11} & \cdots & q_{1r} \\ q_{11}+q_{01} & q_{11} & q_{11}+q_{01} & \cdots & 0 \\ \vdots & \vdots & \vdots & \ddots & \vdots \\ q_{1r}+q_{0r} & q_{1r} & 0 & \cdots & q_{1r}+q_{0r} \end{bmatrix},$$

where r=k−1 and q lm =E(T i e b lm I{individual$i$in category(l,m)}) qlm=E(TieblmI{individualiin category(l,m)})

$q_{lm} = E(T_i e^{b_{lm}} I\{\text{individual } i \text{ in category } (l,m)\})$. The consistent estimator, $\hat{I}(\hat{\beta})$, of $I(\beta)$ is thus obtained by replacing q_{lm} with $\exp(\hat{b}_{lm})t_{lm}/n$

.

Under model (23), the expected number of false events for a subject in category (l,m) with the time-exposure interval $[0,T]$ is $T\exp(d_{lm})$, for l=0,1 and m=1,⋯,k, where $d_{1k}=a_0+a_1$, $d_{0k}=a_0$, $d_{1m}=a_0+a_1+\gamma_m$ and $d_{0m}=a_0+\gamma_m$ for $1 \le m \le k-1$. The conditional distribution of Y given $A=a$, T, and $Z \simeq (l,m)$ is Binomial (a, p_{lm}), $a\hat{p}_{lm}(1-\hat{p}_{lm})$, where $\hat{p}_{lm} = e^{\hat{b}_{lm}}/(e^{\hat{b}_{lm}} + e^{\hat{d}_{lm}})$, and $\mathrm{Var}(T|A=a, Z \simeq (l,m))$ can be estimated by $v_{a,l,m}=t_{2,a,l,m}/n_{alm}-(t_{alm}/n_{alm})^2$. By (24) and the discussion that follows, $\Sigma_{A1}(\beta)$ can be estimated by

$$\hat{\Sigma}_{A1}(\hat{\beta}) = \sum_{a,l,m} \hat{\rho}(a,l,m)\frac{1-\hat{\rho}^{v}(a,l,m)}{\hat{\rho}^{v}(a,l,m)} G_{lm}\{a\hat{p}_{lm}(1-\hat{p}_{lm}) + v_{a,l,m}\}, \tag{26}$$

where

$\hat{\rho}(a,l,m) = n_{alm}/n$, $\hat{\rho}^{v}(a,l,m) = n^{v}_{alm}/n_{alm}$ and G_{lm} be the value of $G_i = z_i^{\otimes 2}$ when subject i belongs to the category (l,m). Hence the covariance matrix of $\hat{\beta}_{I1}$ can be estimated by $\hat{I}^{-1}(\hat{\beta}) + \hat{I}^{-1}(\hat{\beta})\hat{\Sigma}_{A1}(\hat{\beta})\hat{I}^{-1}(\hat{\beta})$ using the automated data.

Estimation with $\hat{\beta}_{A2}$ using the automated data. The estimating equations (25) are equivalent to the following nonlinear equations for $\{b_{lm}$, for l=0,1, m=1,⋯,$k\}$,

$$\sum_{m=1}^{k}\left\{\sum_{a\in A} y_{alm}\lambda_{alm} - e^{b_{lm}}t_{lm} + \frac{e^{b_{lm}}}{e^{b_{lm}}+e^{\hat{d}_{1m}}}\left(\alpha_{lm} - \sum_{a\in A} a n_{alm}\lambda_{alm}\right)\right\} = 0,$$

$$\sum_{l=0,1}\left\{\sum_{a\in A} y_{alm}\lambda_{alm} - e^{b_{lm}}t_{lm} + \frac{e^{b_{lm}}}{e^{b_{lm}}+e^{\hat{d}_{lm}}}\left(\alpha_{lm} - \sum_{a\in A} a n_{alm}\lambda_{alm}\right)\right\} = 0,$$

for l=0,1 and m=1,...,k−1.

Since Var{$S_\beta(Y|X)|Z\simeq(l,m),T,A=a$}=$a\, p_{lm}(1-p_{lm})G_{lm}$, $\Sigma_{A2}(\beta)$ can be consistently estimated by

$$\hat{\Sigma}_{A2}(\hat{\beta}) = \sum_{a,l,m} \hat{\rho}(a,l,m)\frac{1-\hat{\rho}^{v}(a,l,m)}{\hat{\rho}^{v}(a,l,m)} a\hat{p}_{lm}(1-\hat{p}_{lm})G_{lm}.$$

Hence the covariance matrix of $\hat{\beta}_{A2}$ β̂A2 $\hat{\beta}_{A2}$ can be estimated by

$$\hat{I}^{-1}(\hat{\beta}) + \hat{I}^{-1}(\hat{\beta})\hat{\Sigma}_{A2}(\hat{\beta})\hat{I}^{-1}(\hat{\beta})$$

using the automated data.

Remark In the special case where $\rho(\alpha,l,m)\approx 0$ for $\alpha\geq 2$, a much simpler formula for the variance estimator of the log relative risk can be derived. For example in the vaccine safety study, the adverse-event rate is very small. Let

$$w_m = \frac{\alpha_{0m}\alpha_{1m}y_{0m}y_{1m}}{\alpha_{0m}y_{0m}n^{v}_{1m} + \alpha_{1m}y_{1m}n^{v}_{0m}} \Big/ \sum_{m=1}^{k} \frac{\alpha_{0m}\alpha_{1m}y_{0m}n^{v}_{1m}}{\alpha_{0m}y_{0m}n^{v}_{1m} + \alpha_{1m}y_{1m}n^{v}_{0m}}.$$

Then an estimate of variance of $\hat{b}_1$ is given by

Var(b ˆ 1) ˆ =∑ k m=1 w 2 m (1y 1m −1n v 1m +1α 1m +1y 0m −1n v 0 m +1α 0m), Var(b̂1)=∑m=1kwm21y1m−1n1mv+1α1m+1y0m−1n0mv+1 α0m,

$$\widehat{\mathrm{Var}(\hat{b}_1)} = \sum_{m=1}^{k} w_m^2 \left(\frac{1}{y_{1m}} - \frac{1}{n^{v}_{1m}} + \frac{1}{\alpha_{1m}} + \frac{1}{y_{0m}} - \frac{1}{n^{v}_{0m}} + \frac{1}{\alpha_{0m}}\right), \quad (27)$$

which is the weighted sum of the estimated variances for the estimated log relative rate of the exposed versus the unexposed over k groups. The details of deviation are given in the Appendix B.

5. A SIMULATION STUDY

We conduct a simulation study to examine the finite sample performance of the IPW estimator $\hat{\beta}_{I1}$ and the AIPW estimator $\hat{\beta}_{A2}$. We consider the Poisson regression model (22). The covariates Z_1 and Z_2 are generated from the Bernoulli distributions with the probability of success equals to 0.4 and 0.5 respectively. The exposure time T is generated from a uniform distribution on [0,10]. Given $Z=(Z_1,Z_2)$ and T, the outcome variable Y follows a Poisson distribution with mean $T\exp(b_0+b_1Z_1+\vartheta Z_2)$ where $b_0=-0.5$, $b_1=-0.8$ and $\vartheta=-0.6$, and W follows a Poisson distribution with mean $T\exp(a_0+a_1Z_1+\gamma Z_2)$ where $a_0=-1.3$, $a_1=-1.1$, $\gamma=-1$. We set $A=Y+W$.

Four models for the validation sample are considered. Under Model 1, the validation sample is a simple random sample with probability $\pi_i=0.4$. Model 2 considers $\pi_i=0.6$. In Model 3, the validation probability only depends on A through the logistic regression model $\text{logit}\{\pi_i(X,A)\}=A-0.5$ where $X=(Z,T)$. In Model 4, the validation probability depends on A and Z_1 through the logistic regression model $\text{logit}\{\pi_i(X,A)\}=A-Z_1-0.5$.

Tables 1 and 2 present the simulation results for n=50,100,300,500 and 800. Each entry of the tables is based on 1000 simulation runs. Tables 1 and 2 summarize the bias (Bias), the empirical standard error (SSE), the average of the estimated standard error (ESE), and the empirical coverage probability (CP) of 95% confidence intervals of $\hat{\beta}_{I1}$ and $\hat{\beta}_{A2}$ for $\beta=(b_0,b_1,\vartheta)$. We also compare the performance of the estimators $\hat{\beta}_{I1}$ and $\hat{\beta}_{A2}$ with the complete-case (CC) estimator $\hat{\beta}_C$ obtained by simply deleting subjects with missing values of Y_i. As a gold standard, we present the estimation results for the full data where all the values of Y_i are fully observed. Table 1 presents the results under Model 1 and 2, and Table 2 shows the results under Model 3 and 4.

Table 1. Simulation comparison of the IPW estimator $\hat{\beta}_{I1}$, the AIPW estimator $\hat{\beta}_{A2}$ and the complete-case (CC) estimator $\hat{\beta}_{C}$ under various sample sizes and selection probabilities

		b_0				b_1				θ			
n		Bias	SSE	ESE	CP	Bias	SSE	ESE	CP	Bias	SSE	ESE	CP
						Model 1: $\pi_i = .4$							
50	IPW	-.0415	.3561	.1839	.851	-.2175	1.6737	.3354	.864	-.1610	1.2201	.2962	.847
	AIPW	-.0110	.2213	.1664	.890	-.0062	.3099	.2873	.943	-.0186	.3076	.2551	.929
	CC	-.0246	.3398	.2738	.938	-.1515	1.6082	.4730	.968	-.1038	1.1709	.4187	.959
100	IPW	-.0650	.1815	.1404	.870	-.0548	.3120	.2458	.891	-.0249	.2653	.2161	.898
	AIPW	-.0094	.1376	.1187	.914	-.0024	.2284	.1988	.926	.0027	.1994	.1780	.925
	CC	-.0240	.1728	.1685	.948	-.0086	.3086	.2961	.960	.0031	.2556	.2581	.946
300	IPW	-.0368	.0936	.0874	.931	-.0209	.1535	.1460	.946	-.0022	.1419	.1286	.929
	AIPW	-.0027	.0732	.0712	.946	-.0028	.1233	.1165	.940	.0005	.1130	.1046	.938
	CC	-.0092	.0919	.0935	.960	-.0012	.1627	.1634	.952	.0040	.1438	.1432	.952
500	IPW	-.0183	.0698	.0671	.938	-.0172	.1159	.1128	.943	-.0083	.1069	.0993	.933
	AIPW	.0022	.0566	.0550	.936	-.0022	.0956	.0902	.943	-.0068	.0867	.0811	.930
	CC	.0006	.0704	.0716	.949	-.0059	.1268	.1255	.949	-.0046	.1135	.1103	.942
800	IPW	-.0126	.0538	.0527	.942	-.0134	.0862	.0889	.950	-.0029	.0759	.0779	.947
	AIPW	.0011	.0433	.0435	.952	-.0047	.0720	.0713	.956	-.0020	.0638	.0640	.951
	CC	.0002	.0562	.0565	.948	-.0051	.0974	.0990	.958	-.0013	.0844	.0869	.958
						Model 2: $\pi_i = .6$							
50	IPW	-.0316	.2079	.1714	.926	-.0934	.8426	.3112	.944	-.0563	.3320	.2690	.937
	AIPW	-.0072	.1723	.1591	.941	-.0105	.2893	.2772	.950	-.0172	.2653	.2440	.948
	CC	-.0126	.1973	.1949	.959	-.0594	.8369	.3512	.967	-.0278	.3213	.3044	.959
100	IPW	-.0366	.1399	.1259	.926	-.0420	.2363	.2192	.944	-.0100	.2103	.1911	.925
	AIPW	-.0121	.1206	.1133	.941	-.0107	.2069	.1921	.944	.0078	.1764	.1700	.940
	CC	-.0142	.1370	.1345	.947	-.0216	.2379	.2370	.961	.0030	.2103	.2072	.949
300	IPW	-.0138	.0742	.0728	.944	-.0194	.1267	.1250	.957	-.0049	.1064	.1096	.964
	AIPW	-.0030	.0650	.0651	.948	-.0044	.1136	.1093	.949	.0005	.0960	.0974	.956
	CC	-.0017	.0763	.0759	.946	-.0118	.1345	.1328	.951	-.0035	.1147	.1169	.957
500	IPW	-.0069	.0571	.0555	.945	-.0096	.0946	.0965	.947	-.0094	.0866	.0844	.953
	AIPW	.0029	.0495	.0496	.942	-.0032	.0856	.0841	.947	-.0076	.0757	.0749	.955
	CC	.0013	.0577	.0581	.947	-.0034	.1024	.1019	.949	-.0086	.0906	.0899	.954
800	IPW	-.0072	.0437	.0438	.954	-.0069	.0754	.0763	.956	-.0025	.0692	.0664	.947
	AIPW	-.0011	.0401	.0393	.951	-.0019	.0693	.0665	.943	-.0015	.0626	.0590	.931
	CC	-.0012	.0452	.0460	.958	-.0026	.0805	.0806	.952	-.0024	.0723	.0709	.952
						Full data: $\pi_i = 1$							
50		-.0079	.1510	.1466	.952	-.0182	.2691	.2618	.948	-.0104	.2264	.2263	.957
100		-.0079	.1068	.1024	.943	-.0075	.1841	.1798	.948	-.0039	.1560	.1577	.949
300		-.0019	.0596	.0583	.950	-.0081	.1032	.1023	.926	.0001	.0934	.0898	.932
500		.0006	.0452	.0450	.951	-.0041	.0783	.0788	.950	.0014	.0656	.0693	.960
800		-.0004	.0343	.0356	.951	.0025	.0612	.0622	.938	-.0006	.0532	.0547	.955

Table 2. Simulation comparison of the IPW estimator $\hat{\beta}_{I1}$, the AIPW estimator $\hat{\beta}_{A2}$ and the complete-case (CC) estimator $\hat{\beta}_{C}$ under various sample sizes and selection probabilities

n		b_0 Bias	b_0 SSE	b_0 ESE	b_0 CP	b_1 Bias	b_1 SSE	b_1 ESE	b_1 CP	θ Bias	θ SSE	θ ESE	θ CP
						Model 3							
50	IPW	.0081	.1609	.1502	.949	-.0034	.5535	.2790	.954	-.0116	.2592	.2400	.963
	AIPW	-.0070	.1543	.1486	.950	-.0134	.2715	.2690	.958	-.0185	.2364	.2330	.955
	CC	.0230	.1529	.1504	.938	.0798	.5441	.2835	.940	.0648	.2367	.2414	.952
100	IPW	-.0052	.1145	.1077	.948	-.0001	.2073	.2014	.959	.0030	.1789	.1724	.948
	AIPW	-.0124	.1077	.1041	.947	-.0085	.1869	.1840	.957	.0050	.1636	.1606	.948
	CC	.0221	.1074	.1054	.939	.1023	.1830	.1937	.924	.0828	.1625	.1664	.915
300	IPW	-.0011	.0617	.0614	.951	-.0044	.1176	.1157	.952	.0019	.1009	.0993	.953
	AIPW	-.0023	.0582	.0588	.956	-.0051	.1056	.1036	.954	.0022	.0936	.0910	.944
	CC	.0295	.0577	.0596	.924	.1051	.1070	.1095	.824	.0823	.0925	.0946	.861
500	IPW	.0018	.0451	.0473	.958	-.0037	.0853	.0895	.958	-.0069	.0765	.0767	.945
	AIPW	.0009	.0430	.0452	.957	-.0032	.0793	.0798	.947	-.0066	.0689	.0702	.951
	CC	.0317	.0429	.0459	.903	.1077	.0788	.0844	.763	.0737	.0704	.0730	.839
800	IPW	-.0006	.0374	.0375	.951	-.0030	.0671	.0708	.962	.0004	.0617	.0605	.946
	AIPW	-.0003	.0362	.0358	.949	-.0031	.0623	.0631	.954	-.0012	.0577	.0554	.935
	CC	.0315	.0353	.0364	.863	.1065	.0630	.0667	.633	.0786	.0568	.0576	.721
						Model 4							
50	IPW	.0053	.1627	.1504	.948	.0825	.3531	.2832	.913	-.0057	.2736	.2405	.948
	AIPW	-.0085	.1549	.1489	.950	-.0122	.2746	.2752	.966	-.0138	.2395	.2340	.961
	CC	.2295	.2640	.0855	.531	.4513	.3805	.1760	.517	.2954	.3285	.1409	.536
100	IPW	-.0050	.1168	.1085	.939	.0481	.2350	.2130	.922	.0016	.1884	.1761	.940
	AIPW	-.0130	.1083	.1043	.943	-.0067	.1920	.1885	.950	.0066	.1648	.1613	.949
	CC	.0196	.1077	.1063	.943	.2010	.1946	.2087	.820	.0900	.1645	.1702	.910
300	IPW	-.0001	.0630	.0624	.945	-.0001	.1323	.1311	.955	-.0011	.1043	.1038	.946
	AIPW	-.0020	.0588	.0588	.951	-.0052	.1095	.1059	.952	.0012	.0950	.0913	.931
	CC	.0271	.0582	.0601	.930	.2020	.1060	.1173	.576	.0894	.0939	.0965	.863
500	IPW	.0012	.0457	.0480	.951	-.0007	.0966	.1010	.966	-.0054	.0801	.0799	.948
	AIPW	.0006	.0433	.0453	.956	-.0010	.0821	.0813	.941	-.0059	.0697	.0704	.950
	CC	.0291	.0434	.0463	.912	.2047	.0817	.0903	.364	.0815	.0711	.0745	.820
800	IPW	-.0006	.0381	.0380	.949	.0004	.0761	.0794	.967	.0000	.0636	.0630	.947
	AIPW	-.0002	.0362	.0359	.949	-.0016	.0640	.0641	.955	-.0014	.0574	.0555	.942
	CC	.0288	.0356	.0367	.885	.2039	.0644	.0714	.166	.0864	.0570	.0588	.673

Table 1 shows that under Model 1 and Model 2, the bias of all estimators is very small at a level comparable with that of the full data estimator. The bias decreases with increased sample size and the increased level of the validation probability. The empirical standard errors are in good agreement with the corresponding estimated standard errors, except for the IPW estimator when $n \le 100$ and $\pi \le 0.6$. Among them, AIPW has the smallest standard errors for all parameters and sample sizes concerned. The coverage probabilities of the confidence intervals for b_0, b_1 and ϑ are close to the nominal level 95%. When the sample size and the validation probability are both small, for example, n=50 and π=0.4, the IPW has large bias and is unstable but the AIPW still performs well.

Table 2 gives the results under Model 3 and Model 4. The bias remains small for $\hat{\beta}_{I1}$ and $\hat{\beta}_{A2}$. The empirical standard errors are also close to the corresponding estimated standard errors. The coverage probabilities remain close to the nominal level 95% for all IPW and AIPW estimators. However, the complete-case estimator yields larger bias and incorrect coverage probability because of the association between the validation probability and the auxiliary variable A and/or the covariate Z_1, in which case the missing is not missing completely at random. The AIPW performs better than IPW with smaller standard errors.

6. AN APPLICATION

A community-based, nonrandomized, open-label influenza vaccine (CAIV-T) study was conducted in Temple-Belton, Texas during the 2000-2001 influenza season. The total 11,606 healthy children aged 18 months - 18 years were involved in this study and about 20% of them received a single dose of CAIV-T in 2000. The primary clinical outcome was based on an nonspecific case definition called medically attended acute respiratory infection (MAARI), which included all International Classification of Diseases, Ninth Revision, Clinical Modification diagnoses codes (ICD-9 codes 381-383, 460-487) for upper and lower respiratory tract infections, otitis media and sinusitis. MAARI outcomes and demographic data were extracted from the Scott & White Health Plan administrative database. For each visit, one or two International Classification of Diseases, Ninth Revision, Clinical Modification diagnosis codes were listed. Visits for which asthma diagnosis codes alone were noted, without another MAARI code, were excluded. More details about this study can be found in Halloran et al. ([2003]).

Any children representing with history of fever and any respiratory illness were eligible to have a throat swab for influenza virus culture. The decision to obtain specimens was made irrespective of whether a patient had received CAIV-T. The specific case definition was culture-confirmed influenza. Table 3 taken from Halloran et al. ([2003]) contains information on the number of children in three age groups, the number of children who are vaccinated versus unvaccinated, the number of nonspecific MAARI cases, the number of cultures performed, and the number of cultures positive for each group.

Table 3. Study data for influenza epidemic season 2000-01, by age and vaccine group (from Halloran et al.**[2003]**)

Age group (years)	Vaccine	No. of children	No. of MAARI cases	No. of MAARI cases cultured	No. of positive cultures
1.5-4	CAIV-T	537	389	16	0
	None	1844	1665	86	24
5-9	CAIV-T	807	316	17	2
	None	2232	1156	118	53
10-18	CAIV-T	937	219	19	3
	None	5249	1421	123	56
Total	CAIV-T	2281	924	52	5
	None	9325	4242	327	133

With the method developed in Section 4 for Poisson regression, we compare the risk of developing MAARI for children who received CAIV-T to the risk for children who had never received CAIV-T using the automated information provided in Table 3. The number of nonspecific MAARI cases extracted using the ICD-9 codes is the auxiliary outcome *A*, whereas the actual number of influenza cases *Y* is the outcome of interest. Let Z_1 be the treatment indicator (1=vaccine and 0=placebo). Let $Z_2=(\eta_1,\eta_2)$ be the dummy variables indicating three age groups, where $\eta_1=1$ if the age is in the range 1.5–4, $\eta_1=0$, otherwise, and $\eta_2=1$ if the age is in the range 5–9, $\eta_2=0$, otherwise. The reference group is the age 10–18. The exposure time for all children is taken as *T*=1 year.

Consider a Poisson regression model with mean *T* $\exp(b_0+b_1Z_1+\vartheta_1\eta_1+\vartheta_2\eta_2)$. Using the IPW estimator β^ I1 $\hat{\beta}_{I1}$, the estimates (standard errors) are $\hat{b}_0=-0.7659$ $\hat{\sigma}_{b_0}=0.1046$), $\hat{b}_1=-1.5830$ $\hat{\sigma}_{b_1}=0.5017$), $\hat{\theta}_1=-0.5572$ $\hat{\sigma}_{\theta_1}=0.2111$) and $\hat{\theta}_2=-0.0199$ $\hat{\sigma}_{\theta_2}=0.1472$). The age-adjusted relative rate (RR) in the vaccinated group compared with the unvaccinated group equals exp $\exp(\hat{b}_1)=\exp(-1.5830)=0.2054$, which means that the rate of developing MAARI for the vaccinated group is 20% of that for the unvaccinated group. In terms of the vaccine efficacy VE=1−RR=0.7946, this represents about 80% reduction in the risk of developing MAARI for the vaccinated group compared to the unvaccinated group. The 95% confidence interval of RR obtained by using the delta method is (0.0768,0.5490), showing clear evidence that the vaccinated children have less risk of influenza than the unvaccinated children. The 95% confidence interval for VE is (0.4510,0.9232).

Using the AIPW estimator $\hat{\beta}_{A2}$, the estimates (standard errors) are $\hat{b}_0=-2.0703$ $\hat{\sigma}_{b_0}=0.0851$, $\hat{b}_1=-1.8072$ $\hat{\sigma}_{b_1}=0.3786$), $\hat{\theta}_1=0.6452$

$\hat{\sigma}_{\theta_1} = 0.1966$) and $\hat{\theta}_2 = 0.6235$ $\hat{\sigma}_{\theta_2} = 0.1265$). The age-adjusted relative rate (RR) is $\exp(\hat{b}_1) = \exp(-1.8072) = 0.1641$. The estimated VE is 0.8359 and the 95% confidence interval is (0.6553,0.9219). The estimator $\hat{\beta}_{A2}$ yields smaller standard errors and confidence intervals with more precision than using $\hat{\beta}_{I1}$.

This data was analyzed by Halloran et al. ([2003]) and Chu and Halloran ([2004]). Assuming the binary probability model for $P_\theta(Y|X)$ where X includes the vaccination status and age group indicators, and using the mean score method, Halloran et al. ([2003]) found that the estimated VE based on the nonspecific MAARI cases alone was 0.18 with 95% confidence interval of (0.11,0.24). The estimated VE by incorporating the surveillance cultures was 0.79 with 95% confidence interval of (0.51,0.91). Halloran et al. also reported sample-size-weighted VE =0.77 with 95% confidence interval of (0.48,0.90). Chu and Halloran ([2004]) have developed a Bayesian method to estimate vaccine efficacy. By Chu and Halloran ([2004]), the estimated VE was 0.74 with 95% confidence interval (0.50,0.88) and estimated VE by the multiple imputation method was 0.71 with 95% confidence interval (0.42,0.86).

Our estimates of the vaccine efficacy are in line with the existing methods. The estimator $\hat{\beta}_{A2}$ yields smaller standard errors and therefore confidence intervals are more precise than the existing methods of Halloran et al. ([2003]) and Chu and Halloran ([2004]). Compared to the binary regression, Poisson regression model allows multiple recurrent MAARI cases for each child. Although for this particular application the exposure time is fixed at one year time interval, the proposed method is applicable to the situation where the length of exposure time may be different for different children.

7. CONCLUSIONS

In this paper, we investigated the mean score method, the IPW method and the AIPW method for the parametric probability regression model $P_\beta(Y|X)$ when outcome of interest Y is subject to missingness. The asymptotic distributions are derived for the IPW estimator and the AIPW estimator. The selection probability often needs to be estimated for the IPW estimator, and both the selection probability and the conditional expectation of the score function needs to be estimated for the AIPW estimator. We investigated the properties of the IPW estimator and the AIPW estimator when the selection probability and the conditional expectation are implemented differently.

An AIPW estimator is said to be fully augmented if the selection probability and the conditional expectation are estimated using the full set of observed variables; it is partially augmented if the selection probability and the conditional expectation are estimated using a subset of observed variables. Corollary 1 shows that the fully augmented AIPW estimator is more efficient than the partially augmented AIPW estimator. Corollary 2 shows that the AIPW estimator is more efficient than the IPW estimator. However, when the selection probability depends only on a set of discrete random variables, the IPW estimator obtained by estimating the selection probability nonparametrically with the cell frequencies is asymptotically equivalent to the AIPW estimator augmented using the same set of discrete random variables. Proposition 1 shows that the IPW estimator, the AIPW estimator and the mean score estimator are equivalent if the selection probability and the conditional expectation are estimated using same set of discrete random variables.

Applying the developed missing data methods, we derived the estimation procedures for Poisson regression model with missing outcomes based on auxiliary outcomes and a validated sample for true outcomes. By assuming the selection probability depending only on the observed discrete exposure variables, not on the continuous exposure time, we show that the IPW estimator and the AIPW estimator can be formulated to analyze data when only aggregated/summarized information are available. The simulation study shows that for a moderate sample size and selection probability, the IPW estimator and AIPW estimator perform better than the complete-case estimator. The AIPW estimator is more efficient and more stable than the IPW estimator. The proposed methods are applied to analyze a data set from for an influenza vaccine study conducted in Temple-Belton, Texas during the 2000-2001 influenza season. The data set presented in Table 3 only contains summarized information at categorical levels defined by the three age groups and vaccination status. The actual number of influenza cases (the number of positive cultures) out of the number of MAARI cases cultured, along with the number of MAARI cases, are available for each category. Our analysis using the AIPW approach shows that the age-adjusted relative rate in the vaccinated group compared to the unvaccinated group equals 0.1641, which represents about 84% reduction in the risk of developing MAARI for the vaccinated group compared to the unvaccinated group.

ACKNOWLEDGEMENTS

The authors thank the two referees for their valuable comments. The authors also thank Yuriko Nagano for some help with the data analysis. This research was partially supported by the National Science Foundation grant DMS-1208978, the National Institute of Health NIAID grant R37 AI054165, and a fund provided by The University of North Carolina at Charlotte.

AUTHORS' CONTRIBUTIONS

YS proved Proposition 1, Theorem 1, Corollary 1 and 2, and wrote the manuscript. LQ developed the Poisson regression using the automated data with missing outcomes, performed the computations for the simulation study and for the application, and wrote the manuscript. Both authors read and approved the final manuscript.

REFERENCES

1. Carpenter JR, Kenward MG, Vansteelandt S: A comparison of multiple imputation and doubly robust estimation for analyses with missing data. J. R. Stat. Soc. A 2006, 169: 571–584. 10.1111/j.1467-985X.2006.00407.x

2. Clayton D, Spiegelhalter D, Dunn G, Pickles A: Analysis of longitudinal binary data from multiphase sampling (with discussion). J. R. Stat. Soc. B 1998, 60: 71–87. 10.1111/1467-9868.00109

3. Chu H, Halloran EM: Estimating vaccine efficacy using auxiliary outcome data and a small validation sample. Stat. Med 2004, 23: 2697–2711. 10.1002/sim.1849

4. Dempster AP, Laird NM, Rubin DB: Maximum likelihood from incomplete data via the EM algorithm. J. R. Stat. Soc. B 1977, 39: 1–38.

5. Ellenberg SS, Hamilton JM: Surrogate endpoints in clinical trials: cancer. Stat. Med 1989, 8: 405–413. 10.1002/sim.4780080404

6. Fleming TR: Evaluating therapeutic interventions: some issues and experiences. Stat. Sci 1992, 7: 428–456. 10.1214/ss/1177011128

7. Halloran EM, Longini IM, Gaglani MJ, Piedra PA, Chu H, Herschler GB, Glezed WP: Estimating efficacy of trivalent, cold-adapted, influenza virus vaccine (CAIV-T) against influenza A (H1N1) and B Using Surveillance Cultures. Am. J. Epid 2003, 158: 305–311. 10.1093/aje/kwg163

8. Horvitz DG, Thompson DJ: A generalization of sampling without replacement from a finite universe. J. Am. Stat. Assoc 1952, 47: 663–685. 10.1080/01621459.1952.10483446

9. Little RJA, Rubin DB: Statistical analysis of missing data. Wiley, New York; 2002.

10. Pepe MS, Reilly M, Fleming TR: Auxiliary outcome data and the mean score method. J. Stat. Plan. Inference 1994, 42: 137–160. 10.1016/0378-3758(94)90194-5

11. Prentice RL: Surrogate endpoints in clinical trials: definition and operational criteria. Stat. Med 1989, 8: 431–440. 10.1002/sim.4780080407

12. Rubin DB: Multiple imputation for nonresponse in surveys. Wiley, New York; 1987.

13. Robins JM, Rotnitzky A, Zhao LP: Estimation of regression coefficients when some regressors are not always observed. J. Am. Stat. Assoc 1994, 89: 846–866. 10.1080/01621459.1994.10476818

14. Scharfstein DO, Rotnitzky A, Robins JM: Adjusting for nonignorable drop-out using semiparametric nonresponse models: rejoinder. J. Am. Stat. Assoc 1999, 94: 1135–1146.

Chapter 7

Statistical method for performance analysis of WBAN in time-saturation

Pham Thanh Hiep[1,2]

Yokohama National University

Le Quy Don Technical University

ABSTRACT

Performance analysis of carrier sense multiple access with collision avoidance (CSMA/CA) scheme based on IEEE 802.15.6 standard has been discussed in many literatures. In previous works on IEEE 802.15.6 and another standards, the performance was analyzed by using access probability in one slot duration and the Markov chain model was proposed to calculate the access probability in both saturation and non-saturation models. However, for both saturation and non-saturation models, the access probability or the packet arrival rate was assumed to be fixed and the effect of remained packets that was transmitted unsuccessfully due to busy channel or collision was not considered. In this paper, in order to evaluate the system performance more accurately, the effect of the remained packet is taken into account and a statistical method is proposed to calculate the successful probability of wireless body area networks (WBANs). Moreover, a time-saturation model is defined as a system model in which the system performance is changed following the operating time, and then, the change between three models, i.e., saturation, non-saturation, and time-saturation models, is analyzed. The calculation result indicates that the proposal method can achieve the same performance as the Markov chain method in both saturation and non-saturation models. However, in the time-saturation model, the proposal statistical method is more accurate.

KEYWORDS

CSMA/CA based on IEEE 802.15.6 standard Saturation model Non-saturation model Time-saturation model Performance analysis of WBAN

1 INTRODUCTION

The aging society, the limited health care resources, the miniaturization of biomedical sensors, and a wide application market have triggered the concept of wireless body area networks (WBANs) and received considerable attentions in the academy and industry. The WBAN applications contain medical applications, entertainment applications, and disability-assistance applications. The IEEE 802.15 has established Task Group 6 (TG6) for an appropriate communication standard for WBANs, called IEEE 802.15.6. The first version of the standard had been issued in February 2012 [1].

An efficient media access control (MAC) protocol is very important to guarantee the timely delivery of emergency traffic. Therefore, the analysis of IEEE 802.15.6 is necessary, and there have been some works on MAC protocol. The efficient MAC protocols to improve the energy consumption were proposed in [2, 3]. Comprehensive investigations of WBAN were represented in [4, 5]. For in-body communications, RF technologies and MAC protocols were introduced in [6, 7].

In the IEEE 802.15.6 standard, user priorities (UPs) are divided into eight levers for different applications. A sensor may initiate frame transaction in different contention-based access phases using carrier sense multiple access with collision avoidance (CSMA/CA) and Slotted Aloha. A coordinator of WBANs has the ability of arranging committed scheduled allocation intervals, unscheduled bi-link allocation intervals on the best-effort basis, and improvised polled and posted allocation intervals in contention-free access phases. The detailed overview of MAC and physical (PHY) functionalities in IEEE 802.15.6 was represented in [8–10].

The performance analysis of the CSMA/CA scheme based on IEEE 802.15.6 has been discussed in many literatures. The theoretical maximum throughput and minimum delay limit of IEEE 802.15.6 was descripted for different frequency bands and data rates in [11–13]. Rashwand et al. are focusing on the impact of different access phase lengths to evaluate the network performance [14–17]. The analytical model based on the Markov chain and the throughput performance of IEEE 802.11 were proposed [18]. In [19, 20], authors represented the network performance of different access categories in IEEE 802.11e and IEEE 802.15.4, respectively.

In the previous works on IEEE 802.15.6 and another standard, the performance was analyzed by using the access probability in one slot duration. The Markov chain model was proposed to calculate the access probability in both saturation and non-saturation models. The saturation model is defined as the system model in which there is at least one sensor that has a packet to send, whereas the non-saturation model is the system model that does not always have a sensor that has a packet to send. However, for both saturation and non-saturation models, the access probability and/or the packet arrival rate meaning the number of generated packets per second were assumed to be fixed.

The effect of remained packets that were not transmitted successfully due to busy channel or collision was not considered. It means that the system performance was analyzed in 1 s from the start.

In this paper, the effect of the remained packet is taken into account and the statistical method to calculate a convergence of throughput based on CSMA/CA scheme of IEEE 802.15.6 is proposed. The term of time-saturation model is defined as a system model whose performance is changed following the operating time. In the previous works, the system model was assumed to be the saturation or the non-saturation model and then the performance was analyzed. However, based on the proposal calculation method, we indicate that the successful probability as well as the system throughput converge on values and the time-saturation model can be the saturation or non-saturation model according to the system parameters. Therefore, the proposal method is more flexible and expected to be more accurate. The relation between three models, i.e, saturation, non-saturation, and time-saturation models, is analyzed and the change between these three models based on the system parameters is represented.

The rest of the paper is organized as follows. We introduce the main concept of IEEE 802.15.6 standard in Section 2. Section 3 shows the performance analysis of WBAN by the discrete-time Markov chain method and the proposal method. The numerical evaluation is described in Section 4. Finally, Section 5 concludes the paper.

2 RELATED DESCRIPTION OF THE IEEE 802.15.6 STANDARD

In this section, we describe some parts of the IEEE 802.15.6 standard that relate to our research. The detail of this standard is represented in [1].

2.1 Physical layer

The current IEEE 802.15.6 standard defines three PHY layers, i.e., narrowband (NB), ultra wideband (UWB), and human body communications (HBC) layers. The selection of each PHY depends on requirements of application. In this work, since we focus on performance analysis of WBAN based on CSMA/CA scheme, NB is considered as an example.

The NB PHY is responsible for activation/deactivation of the radio transceiver, clear channel assessment (CCA) within the current channel, and data transmission/reception. The physical protocol data unit (PPDU) frame of NB PHY contains a physical layer convergence procedure (PLCP) preamble, a PLCP header, and a PHY service data unit (PSDU) as given in Figure 1. The PLCP preamble helps the receiver in the timing synchronization and carrier-offset recovery. It is the first component being transmitted at the given symbol rate. The PLCP header conveys information necessary for a successful decoding of a packet to the receiver. The PLCP header is transmitted after the PLCP preamble using the given header data rate in the operating frequency band. The last component of PPDU is PSDU which consists of a MAC header, MAC frame body (also called as payload), and frame check sequence (FCS) and is transmitted after the PLCP header using any of the available data rates in the operating frequency band. A WBAN device should be able to support transmission and reception in one of the frequency bands summarized in Table 1. (Further detail for modulation and channel coding can be found in [1, 10]).

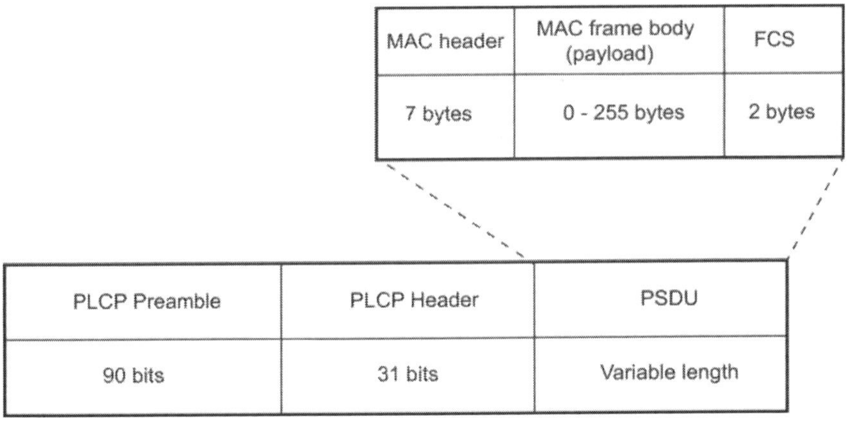

Figure 1. IEEE 802.15.6 PPDU structure.

Table 1. Main parameter for NB

Frequency band	Packet component	Symbol rate (Ksps)	Data rate (Kbps)
420 to 450 MHz	PLCP header	187.5	57.5
	PSDU	187.5	75.9
	PSDU	187.5	151.8
	PSDU	187.5	187.5
863 to 870 MHz	PLCP header	250	76.6
950 to 956 MHz	PSDU	250	101.2
	PSDU	250	202.4
	PSDU	250	404.8
	PSDU	250	607.1
902 to 928 MHz	PLCP header	300	91.9
	PSDU	300	121.4
	PSDU	300	242.9
	PSDU	300	485.7
	PSDU	300	728.6
2360 to 2400 MHz	PLCP header	600	91.9
2400 to 2483.5 MHz	PSDU	600	121.4
	PSDU	600	242.9
	PSDU	600	485.7
	PSDU	600	971.4

2.2 Introduction of CSMA/CA based on IEEE 802.15.6

In IEEE 802.15.6, there are three access mechanisms that are comprehensively discussed in the standard. 1) Random access mechanism, which uses either CSMA/CA or a slotted aloha procedure for resource allocation; 2) Improvised and unscheduled access (connectionless contention-free access), which uses unscheduled polling/posting for resource allocation; and 3) Scheduled access and variants (connection-oriented contention-free access), which schedule the allocation of slots in one or multiple upcoming superframes, also called 1-periodic or m-periodic allocations. Since the CSMA/CA is considered in our analysis, the basic procedures of this protocol is explained as defined in the standard.

In CSMA/CA, a sensor sets its backoff counter to a random integer number uniformly distributed over the interval $[1,W]$ where $W\in(W_{min},W_{max})$ denotes the contention window. The values of W_{min} and W_{max} change depending on the UPs as given in Table 2. The sensor decreases the backoff counter by one for each idle CSMA slot of duration. Particularly, the sensor treats a CSMA slot to be idle if it determines that the channel has been idle between the start of the CSMA slot and *pCCATime*. If the backoff counter reaches zero, the sensor transmits a packet of data. If the channel is busy because of transmission of another sensor, the sensor locks its backoff counter until the channel is idle. The W is doubled for even number of failures until it reaches W_{max}. The failure means that the sensor fails to receive an acknowledgement from the coordinator. Further details about the CSMA/CA procedure can be found in the standard [1].

2.3 Calculation of service time

The service time (T) is defined as the total time to transmit a packet including the *backoff* time (T_W), the time to transmit a data packet (T_{data}), short interframe spacing (T_{pSIFS}), the time of acknowledgement packet (T_{ACK}), and delay time (α) that is defined as the sum of propagation delay and signal processing delay.

$$T = T_W + T_{DATA} + T_{ACK} + 2T_{pSIFS} + 2\alpha. \tag{1}$$

Table 2. Contention window bound for CSMA/CA

User priority	W_{min}	W_{max}
0	16	64
1	16	32
2	8	32
3	8	16
4	4	16
5	4	8
6	2	8
7	1	2

As mentioned above, according to the IEEE 802.15.6 standard, the W is doubled for even number of failures until it reaches W_{max}, hence the value of the average backoff time can be obtained as

$$T_W = \begin{cases} \frac{W_{min} T_s}{2} + P_{fail}^2 W_{min} T_s + P_{fail}^4 \frac{W_{max} T_s}{2} & \text{for even UPs,} \\ \frac{W_{min} T_s}{2} + P_{fail}^2 \frac{W_{max} T_s}{2} & \text{for odd UPs,} \end{cases} \tag{2}$$

here, P_{fail} represents the transmission failed probability and T_s denotes a CSMA slot length.As shown in Figure 1, since a data packet consists of a preamble, physical header, MAC header, MAC frame body, and frame check sequence, the time to transmit a data packet is represented as follows.

$$T_{\mathrm{DATA}} = T_P + T_{\mathrm{PHY}} + T_{\mathrm{MAC}} + T_{\mathrm{BODY}} + T_{\mathrm{FCS}}, \tag{3}$$

here T_P, T_{PHY}, T_{MAC}, T_{BODY}, and T_{FCS} represent the time to transmit a preamble, physical header, MAC header, MAC frame body, and frame check sequence, respectively.

Since an immediate acknowledgement carries no payload, its transmission time is given by

$$T_{\mathrm{ACK}} = T_P + T_{\mathrm{PHY}} + T_{\mathrm{MAC}} + T_{\mathrm{FCS}}. \tag{4}$$

The other parameters to calculate the service time are summarized in Table 3.

Table 3. Other parameters

Clear channel assessment [bits]	63
MAC header [bits]	56
MAC footer [bits]	16
Short interframe spacing time T_{sifs}[μs]	75
Preamble [bits]	90
Delay time τ [μs]	1

3 PERFORMANCE ANALYSIS FOR CSMA/CA BASED ON IEEE 802.15.6

3.1 Previous researches

The discrete-time Markov chain (DTMC) method has been applied to calculate the throughput of WBAN based on IEEE 802.15.6 for both the saturation and the non-saturation models in many literatures [14–17]. Figure 2 shows an example of the discrete-time Markov chain method for the non-saturation model; here ρ, $P_{i,\mathrm{fail}}$, $P_{i,\mathrm{idle}}$, and m represent a probability of having a packet to send in a slot time, transmission failed probability, channel idle probability, and retry limit number, respectively.

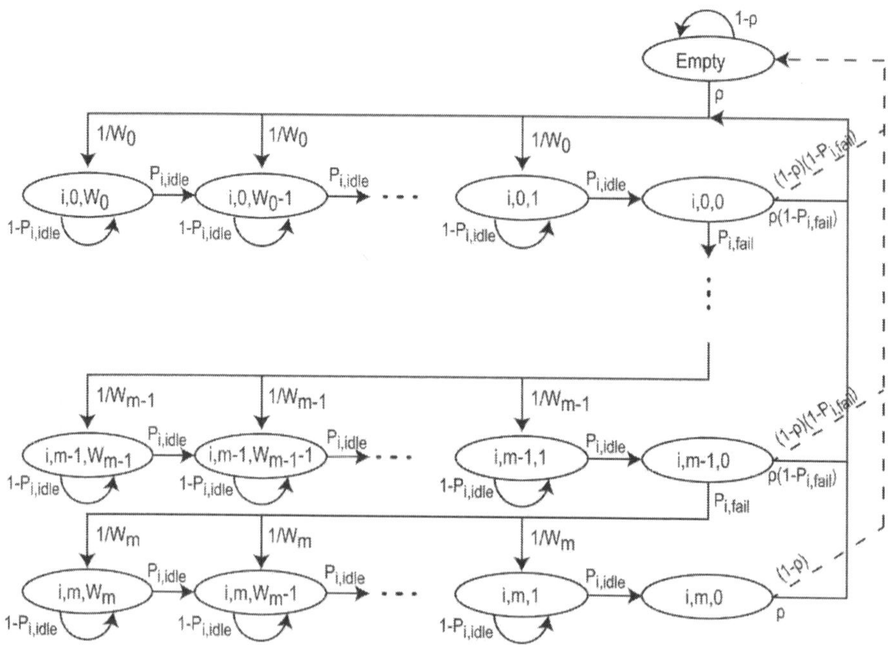

Figure 2. Discrete-time Markov chain method.

By using the state transition probabilities, the stationary distribution $b_{i,0,0}$ and the access probability in one slot duration τ_i are calculated.

$$\tau_i = \frac{1 - P_{i,\text{fail}}^{m+1}}{1 - P_{i,\text{fail}}} b_{i,0,0}. \tag{5}$$

Let P_{tr} and $P_{i,\text{suc}}$ denote the transmission probability and the successful probability, respectively. The transmission probability is defined as the probability that there is at least one sensor that accesses the medium in the given slot. The successful probability is defined as the probability that the data packet is successfully received at the coordinator. Therefore, the throughput of sensors in the ith UP can be represented by

$$S_i = \frac{P_{i,\text{suc}} P_{tr} E[x]}{(1 - P_{tr})T_s + P_{i,\text{suc}} P_{tr} T + (1 - P_{i,\text{suc}}) P_{tr} T_c}, \tag{6}$$

here, T_c and $E[x]$ denote the average time the channel is sensed as busy by each station due to a collision and the average packet payload size, respectively.

However, the probability of having a packet to transmit (the changing of stationary distribution from $b_{i,j,0}$ (j denotes a backoff stage and is bound by a retry limit m, j=0,⋯,m) to $b_{i,0,W}$ (W=1,⋯,W_0)) is considered as ρ. Practically, it should be considered in a duration time in which the sensor has been trying to transmit the generated packet. The probability of having a packet to send is much more than ρ and changes depending on the value of j. In case the probability of having a packet to send is not fixed, the calculation of access probability becomes considerably complicated. Therefore, we propose the statistical method to analyze the performance.

3.2 Proposal method

In our method, the average backoff time is used. Therefore, the average service time *T* is considered and, the performance in each service time is analyzed. All sensors are assumed to be in the same UP. The probability that there is a packet to transmit after *T* duration time is described by

$$\rho = 1 - e^{-\lambda T}, \tag{7}$$

here, λ denotes the packet arrival rate. A packet is successfully transmitted if 1) only one sensor has a packet to send (scheme 1) and 2) more than one sensor has packet to send; however, only one sensor accesses the channel (scheme 2). The probability of scheme 1 is represented as follows.

$$P_{\mathrm{suc}_1} = \binom{N}{1}\rho(1-\rho)^{N-1}, \tag{8}$$

here, *(ji)* ji$\binom{j}{i}$ denotes the binomial coefficient indexed by *j* and *i*; *N* denotes the number of sensors. In scheme 2, the number of sensors that have a packet to send is denoted as *i*; only one sensor successfully transmits when its backoff counter is the smallest and there is not another sensor that has the same backoff counter. As mentioned above, the average backoff counter is used,

$$\overline{W} = \begin{cases} \frac{W_{\min}}{2} + P_{\mathrm{fail}}^2 W_{\min} + P_{\mathrm{fail}}^4 \frac{W_{\max}}{2} & \text{for even UPs,} \\ \frac{W_{\min}}{2} + P_{\mathrm{fail}}^2 \frac{W_{\max}}{2} & \text{for odd UPs.} \end{cases} \tag{9}$$

The smallest backoff counter is assumed to be set as $W \in [1, \overline{W} - 1]$. The backoff counter of other sensors therefore should be higher than *W* meaning it should be within $[W + 1, \overline{W}]$. The probability that the backoff count of a sensor is set as *W* is represented as *(i1)* $\binom{i}{1}\frac{1}{\overline{W}}$, and the probability that the backoff count of the other sensors is within $[W+1, \overline{W}]$ is represented as $\left(\frac{\overline{W}-W}{\overline{W}}\right)^{i-1}$. Therefore, the probability of scheme 2 is described as

$$P_{\mathrm{suc}_2} = \sum_{i=2}^{N} \binom{N}{i} \rho^i (1-\rho)^{N-i} \sum_{W=1}^{\overline{W}-1} \binom{i}{1} \frac{1}{\overline{W}} \left(\frac{\overline{W}-W}{\overline{W}}\right)^{i-1}. \tag{10}$$

Since $0^0=1$ and $0^x=0$ ($x \neq 0$), the successful probability of both schemes 1 and 2 can be unified as

$$\begin{aligned} P_{\mathrm{suc}} &= P_{\mathrm{suc}_1} + P_{\mathrm{suc}_2} \\ &= \sum_{i=1}^{N} \binom{N}{i} \rho^i (1-\rho)^{N-i} \sum_{W=1}^{\overline{W}} \binom{i}{1} \frac{1}{\overline{W}} \left(\frac{\overline{W}-W}{\overline{W}}\right)^{i-1}. \end{aligned} \tag{11}$$

The throughput in each T duration time is calculated as

$$S_T = P_{\mathrm{suc}} E[x], \tag{12}$$

and the throughput of the system in 1 s is described as

$$S = n_T S_T, \tag{13}$$

here, n_T denotes the number of T duration times in each second, $n_T = \mathrm{floor}(\frac{1}{T})$.

However, this method is just used to analyze the performance of the system in 1 s after starting. The throughput in (13) is the same as the average throughput of previous researches. However, as explained in Section 1, the convergence of throughput should be estimated. In order to discuss the effect of the remained packet, the successful probability after t times of T should be considered. The successful probability in (11) is rewrittenas

$$P_{\mathrm{suc}}(t) = \sum_{i=1}^{N} \binom{N}{i} (P_{\mathrm{pac}}(t))^{i} (1 - P_{\mathrm{pac}}(t))^{N-i} \sum_{W=1}^{\overline{W}} \binom{i}{1} \frac{1}{\overline{W}} \left(\frac{\overline{W} - W}{\overline{W}} \right)^{i-1}, \tag{14}$$

here, $P_{\mathrm{pac}}(t)$ denotes the probability that a sensor has a packet to send after t times of T. Notice that

$$\begin{aligned} &P_{\mathrm{pac}}(t) = \rho, \quad \text{for } t = 1. \\ &P_{\mathrm{pac}}(t) = \rho + P_{\mathrm{pac}}(t-1) - \frac{P_{\mathrm{suc}}(t-1)}{N}, \quad \text{for } t > 1. \end{aligned} \tag{15}$$

In the CSMA/CA scheme based on IEEE 802.15.6, the packet is discarded after the retry limit. However, according to the application, the time-out data is deleted; even the number of retransmission does not reach the retry limit. The time-out of data is set as k times of T. It means that at the time $t=k+1$ the packets that were generated at time $t=1$ are discarded. Hence, the probability $P_{\mathrm{pac}}(t)$ is changed as

$$\begin{aligned} &P_{\mathrm{pac}}(t) = \rho, \quad \text{for } t = 1. \\ &P_{\mathrm{pac}}(t) = \rho + P_{\mathrm{pac}}(t-1) - \frac{P_{\mathrm{suc}}(t-1)}{N}, \quad \text{for } 1 < t \leq k, \\ &P_{\mathrm{pac}}(t) = \rho + P_{\mathrm{pac}}(t-1) - \frac{P_{\mathrm{suc}}(t-1)}{N} \\ &\qquad - P_{\mathrm{pac}}(t-k) \prod_{l=t-k}^{t-1} \frac{P_{\mathrm{fail}}(l)}{N}, \quad \text{for } t > k, \end{aligned} \tag{16}$$

here, $P_{\mathrm{fail}}(t)$ denotes the transmission failed probability at time t. A collision occurs when at least two sensors have a packet to send and the backoff counter of at least two sensors is set as the smallest value.

$$P_{\text{fail}}(t) = \sum_{i=2}^{N} \binom{N}{i} \left(P_{\text{pac}}(t)\right)^{i} \left(1 - P_{\text{pac}}(t)\right)^{N-i} \times \left(1 - \sum_{W=1}^{\overline{W}} \binom{i}{1} \frac{1}{\overline{W}} \left(\frac{\overline{W} - W}{\overline{W}}\right)^{i-1}\right). \tag{17}$$

Furthermore, in some scenarios, a sensor may have more than one packet to send and then the probability $P_{\text{pac}}(t)$ is over one. In these scenarios, the data in the newest packet is updated and the data in the other packets is outdated. Therefore, only the newest packet is remained to send; the other packets are deleted as the time-out packet, and then, the probability $P_{\text{pac}}(t)$ is redefined as

$$P_{\text{pac}}(t) = \min(P_{\text{pac}}(t), 1), \quad \text{for all t.} \tag{18}$$

The system throughput after t_n seconds is changed as

$$\begin{aligned} S_T(t) &= P_{\text{suc}}(t) E[x], \\ S(t_n) &= \sum_{t=t_n-n_T}^{t_n} S_T(t). \end{aligned} \tag{19}$$

4 NUMERICAL EVALUATION

The system model is the same as mentioned above. The parameters summarized in Tables 1, 2, and 3 are used, and the calculation method of throughput is given in Figure 3. The algorithm stops when the throughput is convergent meaning the throughput at t_n^{th} second is the same as the throughput at t_{n-1}^{th} second.

```
Initialization: N, λ, CWmin, CWmax and other parameters
Calculation: ρ, T, CW‾ and set Ppac(1)=ρ.
While (the throughput of system isn't convergent).
    For each number of T
        Calculate Ppac(t) by eq. (16) and Ppac(t)=min(Ppac(t),1).
        For each number of sensors that has a packet to send
            For each contention window
                Calculate the successful probability by eq. (14).
            End for
        End for
        Calculate the transmission failed probability by eq. (17).
        Calculate throughput in each T by eq. (19).
    End for
    Calculate throughput in each tn second by eq. (19).
End while
```

Figure 3. The calculation method of system throughput with one user priority.

The bandwidth frequency, the payload, the number of sensors, and W_{min} are fixed as 2,360 to 2,483.5 MHz, 250 bytes, 20, and 16, respectively. The time-out is set as $k=n_T$; it means the packet is discarded after 1 s. The probability of having a packet to send and the probability of having no packet to send as well as the successful probability of each T duration in cases the lambda is fixed as 2 and 10 packets/second are shown in Figures 4 and 5, respectively. The result of the DTMC method and the proposal method is compared in both figures.

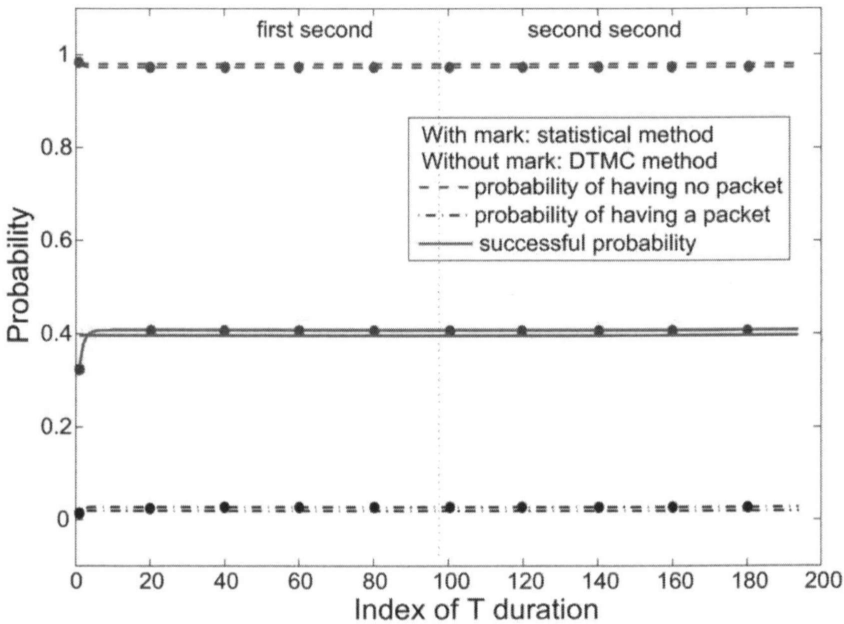

Figure 4. The probability in each **T** duration when the lambda = 2 packets/second.

As shown in Figure 4, when the lambda is small, the probability of having a packet to send is low, and then, the probabilities of having a packet to send and having no packet to send are constant after some *T* durations. Hence, the successful probability is fixed as the total probabilities of having a packet to send. This system model can be considered as the non-saturation model.

On the contrary, when the lambda is large (Figure 5), the probability of having a packet to send reaches 1 after some *T* durations. It means that all sensors always have a packet to send, and the successful probability is fixed as a low value due to the high transmission failed probability. This system model can be considered as the saturation model. Compared to the DTMC method, the result of the proposal method in both the non-saturation (Figure 4) and saturation (Figure 5) models is almost the same.Compare the successful probability of DTMC method in both cases, the lambda is 4 and 10 packets/second. The successful probability when the lambda is 4 packets/second (Figure 6) is higher than that when the lambda is 10 packets/second (Figure 5). The reason is that the DTMC method does not consider the remained packet; therefore the probability of having a packet when the lambda is 10 packets/second is higher, and then, the collision probability is higher. As a result, the successful probability is lower than that when the lambda is 4 packets/second.However, for the proposed method, as shown in Figure 6, when the lambda is 4 packets/second, the probabilities are changed gradually. The probability of having a packet increases slowly and achieves one due to the effect of remained packet. Therefore, the successful probability decreases after increasing and reaches the convergent probability when the probability of having a packet achieves one. It means the system model changes from the non-saturation model to the saturation model; this model is called as the time-saturation model. It is different to the saturation and the non-saturation models; in the time-saturation model, the result of the DTMC method and the proposal method is different. The reason can be explained that since the remained packet is not considered in the DTMC method, the collision probability of DTMC method is low meaning the successful probability is high. However, in fact, the remained packet lets the collision probability increase and the successful probability decrease. Therefore, the proposal method that takes the remained packet into account is considered to be more accurate.Furthermore, a system model can change between the saturation, the non-saturation, and the time-saturation models depending on the system parameters, such as the packet arrival rate, the payload, the retry limit, the number of sensors, the UPs, and so on. Figure 7 shows the change between three models of system in which the lambda is fixed as 10 packets/second and the payload is changed, i.e., 50, 100, and 150 bytes. When the payload is 50 bytes, the system model is the non-saturation model; however, when the payload increases to 150 bytes, the system model changes as the saturation model. Additionally, when the payload is 100 bytes, the system model acts as the time-saturation model. It means that a system can change between three models depending on the payload.

Moreover, the time-saturation model can be considered as a connection model between the non-saturation and the saturation models.

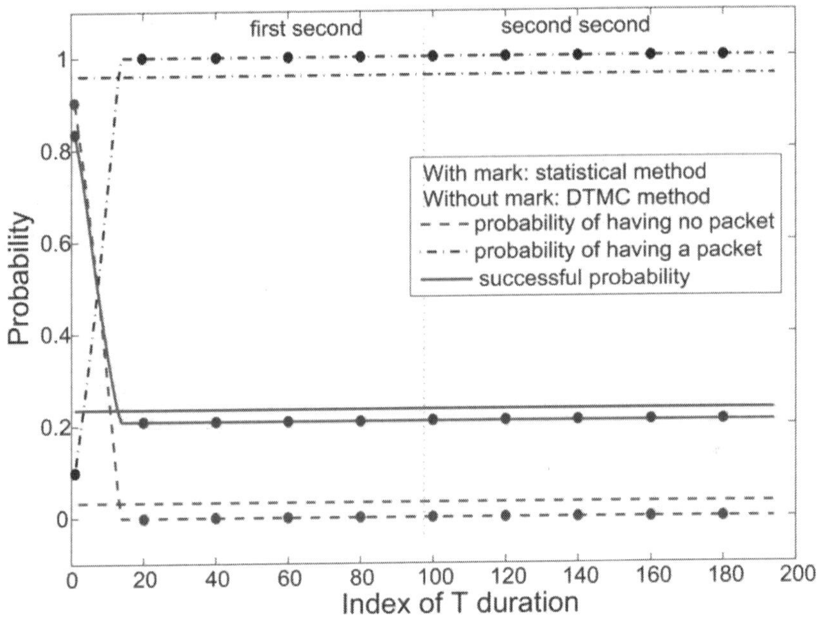

Figure 5. The probability in each **T** duration when the lambda = 10 packets/second.

Figure 8 shows the successful probability when the lambda is fixed as 4 and the *k* is changed, i.e., n_T, 16 and 8. When the *k* reduces, it means that the time-out of packets is short; a system can change from the time-saturation model to the non-saturation model and has the higher successful probability.

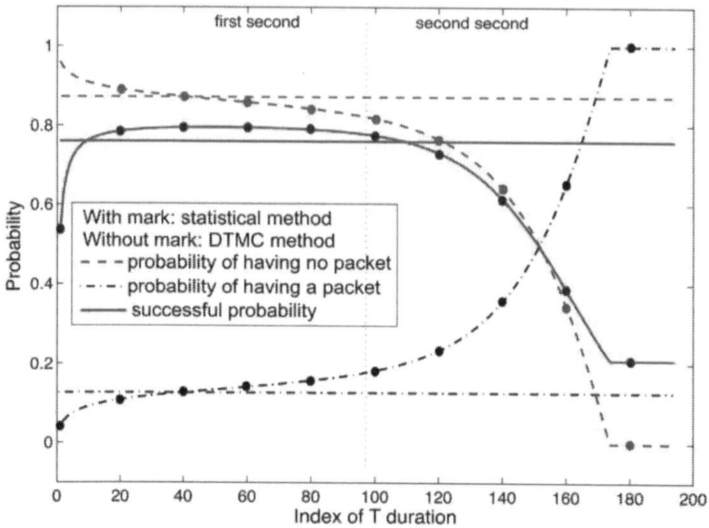

Figure 6. The probability in each **T** duration when the lambda = 4 packets/second.

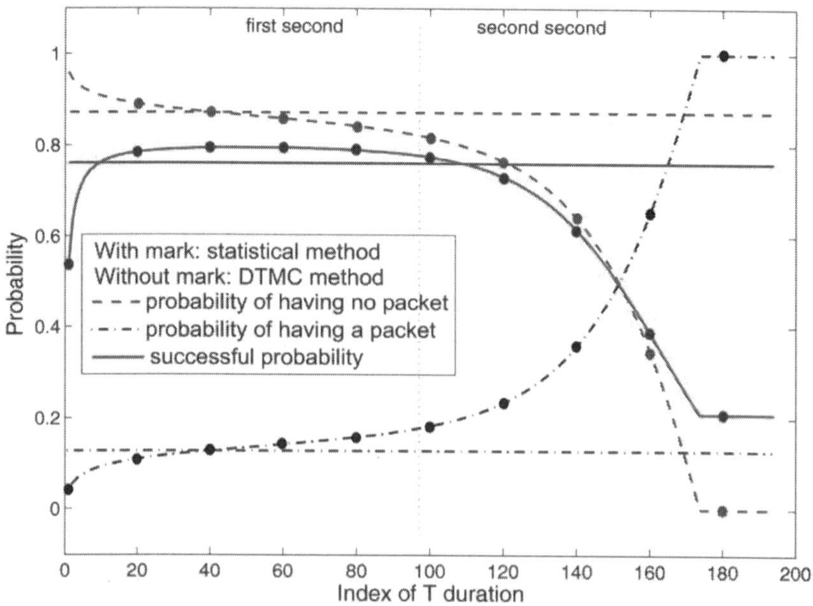

Figure 7. The change between three models when the lambda = 10 packets/section and the payload is changed.

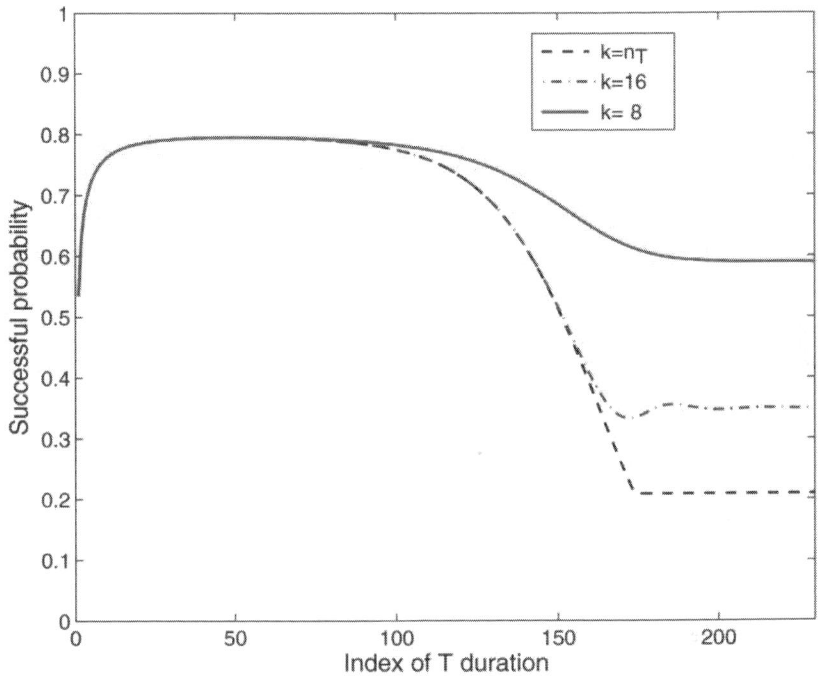

Figure 8. The change between three models when lambda = 4 packets/second and the **k** is changed.

5 CONCLUSIONS

In this paper, the term of time saturation has been defined and the new method has been proposed to analyze system performance in saturation, non-saturation, and time-saturation models. Compared to the DTMC method, the result of the proposal statistical method is more accurate because of taking the remained packet into account. Moreover, the proposal method is more simple because the mathematical calculation is not necessary. The relation between non-saturation, saturation, and time-saturation models was analyzed, and the change between all models depending on the system parameters was described.However, since the performance of non-saturation and saturation system has been analyzed in several literatures, in this paper, another object, such as delay, energy, and so on, was not considered.

Furthermore, we proposed a new method for one user priority; the performance analysis of multiple user priorities is left to the future

REFERENCES

1. WG802.15 - Wireless Personal Area Network (WPAN) Working Group: IEEE standard 802.15.6, wireless body area networks. IEEE Standards 2012, 2: 1-271.

2. Rezvani S, Ghorashi A: A novel WBAN MAC protocol with improved energy consumption and data rate. KSII Trans. Internet Inf. Syst 2012, 6: 2302-2322.

3. Kang Y, Lim S, Kim C: Design, analysis and implementation of energy-efficient broadcast MAC protocols for wireless sensor networks. KSII Trans. Internet Inf. Syst 2011, 5: 1113-1132.

4. Ullah S, Higgins H, Braem B, Latre B, Blondia C, Moerman I, Saleem S, Rahman Z, Kwak KS: A comprehensive survey of wireless body area networks: on PHY, MAC, and network layers solutions. J. Med. Syst 2012, 36: 1065-1094. 10.1007/s10916-010-9571-3

5. Chen M, Gonzalez S, Vasilakos A, Cao H, Leung VCM: Body area networks: a survey. J. Mobile Netw. Appl 2011, 16: 171-193. 10.1007/s11036-010-0260-8

6. Ullah S, Higgins H, Shen B, Kwak KS: Body area networks: a survey. J. Commun. Syst 2010, 23: 982-999.

7. Ullah S, An X, Kwak K: Towards power efficient MAC protocol for in-body and on-body sensor networks. Agent Multi-Agent Syst. Technol. Appl 2009, 5559: 335-345. 10.1007/978-3-642-01665-3_34

8. Bradai N, Belhaj S, Chaari L, Kamoun L: Study of medium access mechanisms under IEEE 802.15.6 standard. In Proceedings of 4th Joint IFIP Wireless and Mobile Networking Conference (WMNC).

University Toulouse III - Paul Sabatier, Toulouse, France, 26–28 Oct; 2011:1-6.

9. Martelli F, Buratti C, Verdone R: On the performance of an IEEE 802.15.6 wireless body area network. In Proceedings of European Wireless 2011, Vienna, Austria. Vienna University of Technology, 27–29 April; 2011.

10. Kwak KS, Ullah S, Ullah N: An overview of IEEE 802.15.6 standard. In Proceedings of 3rd International Symposium on Applied Sciences in Biomedical and Communication Technologies. University of Roma, Rome, Italy, 7–10 Nov.; 2010.

11. Li C, Geng X, Yuan J, Sun T: Performance analysis of IEEE 802.15.6 MAC protocol in beacon mode with superframes. KSII Trans. Internet Inf. Syst 2013, 7: 1108-1130.

12. Ullah S, Chen M, Kwak KS: Throughput and delay analysis of IEEE 802.15.6-based CSMA/CA protocol. J. Med. Syst 2012, 36: 3875-3891. 10.1007/s10916-012-9860-0

13. Ullah S, Kwak KS: Throughput and delay limits of IEEE 802.15.6. In Proceedings of IEEE Wireless Communications and Networking Conference, IEEE Communication Society. Cancun, Quintana Roo, Mexico, 28–31 Mar; 2011:174-178.

14. Rashwand S, Misic J, Khazaei H: Performance analysis of IEEE 802.15.6 under saturation condition and error-prone channel. In Proceedings of IEEE Wireless Communications and Networking Conference, IEEE Communication Society. Cancun, Quintana Roo, Mexico, 28–31 Mar; 2011:1167-1172.

15. Rashwand S, Misic J, Khazaei H: IEEE 802.15.6 under saturation: some problems to be expected. J. Commun. Netw 2011, 13: 142-148.

16. Rashwand S, Misic J: Performance evaluation of IEEE 802.15.6 under non-saturation condition. In Proceedings of IEEE Global

Telecommunications Conference (GLOBECOM), IEEE Communication Society. Texas, USA, 5–9 Dec; 2011.

17. Rashwand S, Misic J: Effects of access phases lengths on performance of IEEE 802.15.6 CSMA/CA. J. Comput. Netw 2012, 56: 2832-2846. 10.1016/j.comnet.2012.04.023

18. Bianchi G: Performance analysis of the IEEE 802.11 distributed coordination function. J. Selected Areas Commun 2000, 18: 535-547.

19. Kong Z, Tsang DHK, Bensaou B, Gao D: Performance analysis of IEEE 802.11e contention-based channel access. J. Selected Areas Commun 2004, 22: 2095-2106. 10.1109/JSAC.2004.836019

20. Li C, Li H, Kohno R: Performance evaluation of IEEE 802.15.4 for wireless body area network (WBAN). In Proceedings of IEEE International Conference on Communications Workshops, IEEE Communication Society. Dresden, Germany, 14–18 June; 2009.

Chapter 8

Change-Point Detection for General Nonparametric Regression Models

Murray D. Burke*, **Gildas Bewa**

Department of Mathematics and Statistics, University of Calgary, Calgary, Canada

Keywords: Change-point detection; Nonparametric regression models; Weighted bootstrap

ABSTRACT

A number of statistical tests are proposed for the purpose of change-point detection in a general nonparametric regression model under mild conditions. New proofs are given to prove the weak convergence of the underlying processes which assume remove the stringent condition of bounded total variation of the regression function and need only second moments. Since many quantities, such as the regression function, the distribution of the covariates and the distribution of the errors, are unspecified, the results are not distribution-free. A weighted bootstrap approach is proposed to approximate the limiting distributions. Results of a simulation study for this paper show good performance for moderate samples sizes.

1. INTRODUCTION

Model building is a very important task. When considering observations $\{(X_i, Y_i): i = 1, 2, \cdots\}$ taken over time we may wish to consider the relationship between the Xi and the Yi by some regression model, either parametric or semiparametric, linear or nonlinear. However, if, after some point i = k, there is a change in the relationship between Xi and Yi, then a single regression model will be inappropriate and would fit the data poorly. For example, in Figure 1 below, the scatterplot on the left is that of a combined sample of 150 points. The middle scatterplot is that of the first 60 points while the right plot is that of the remaining 90 points. Without realizing that there was a change-point, one might wrongly propose a single regression model, perhaps linear, with heteroscedastic errors. Thus, it is important to determine whether such a change has occurred before the particular form of the model is postulated. If such a change has occurred, then two regression models should be used, one for the first k pairs and another for the remaining pairs. We will propose a number of tests designed to detect whether a change has occurred in a general nonparametric model. Change-point problems arise in many areas when observations (Xi, Yi) are taken over time i.

For example, the height of (or discharge from) a river as a function of upstream precipitation, or concentrations of heavy metals in storm-water runoff as a function of suspended sediment size. We refer to the book by Csörgő and Horváth [1] for more applications and important references and to the papers [2-4] for other change-point problems.

We say that the pair (X, Y) satisfies general regression model M(φ, ψ, F, ε) if

$$Y = \phi(X) + \psi(X)\varepsilon, \mathrm{P}[X \le x] = F(x),$$

for all $x \in R^d$, where ε is a mean zero, variance one random variable, independent of X. We write (X, Y) ∈ M(φ, ψ, F, ε).

For a sequence $(X_1, Y_1), \cdots, (X_n, Y_n)$, consider the null hypotheses:

$$(X_i, Y_i) \in M(\phi, \psi, F, \varepsilon), i = 1, \cdots, n, \tag{1}$$

where φ, ψ > 0, and F are unspecified, the Xi are independent and the εi are independent and identically distributed. Model (1) encompasses both linear and nonlinear regression models and the case of heteroscedastic errors. We wish to test whether there is a change in the model as the observations are taken, without specifying φ, ψ or F, that is to test the null Hypothesis (1) versus the alternative:

$$\begin{aligned} &(X_i, Y_i) \in M(\phi, \psi, F, \varepsilon), 1 \le i \le [n\theta], \\ &(X_i, Y_i) \in M(\phi^1, \psi, F, \varepsilon), [n\theta] + 1 \le i \le n, \end{aligned} \tag{2}$$

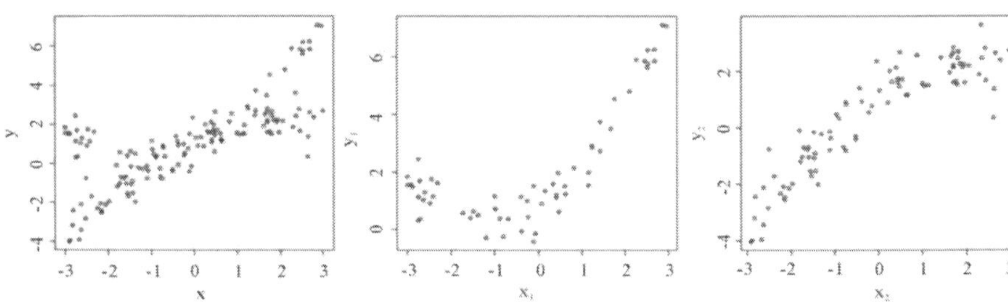

Figure 1. Scatterplot of (Xi, Yi).

for some $\theta \in (0, 1)$, where [s] denotes the integer part of s.

$$\int_{D(x)} \phi \, \mathrm{d}F \neq \int_{D(x)} \phi^1 \mathrm{d}F^1 ;\ D(x) = \{ v \in R^d : v \le x \}, \tag{3}$$

for some $x \in R^d$, using the usual partial order in R^d.

The alternative Hypothesis (2) includes the case where only φ changes, that is, $\phi \neq \phi^1, F = F^1$, for some $\theta \in (0,1)$, and when only F changes $F \neq F^1, \phi = \phi^1$, for some $\theta \in (0,1)$.

We will base our tests on the process

$$\beta_n(x,s) = \frac{1}{\sqrt{n}} \left(\sum_{i=1}^{[ns]} \left(Y I_{D(x)}(X_i) - \overline{Y I_{D(x)}(X)_n} \right) \right)$$
$$= \frac{1}{\sqrt{n}} \left(\sum_{i=1}^{[ns]} \left(Y_i I_{D(x)}(X_i) - \frac{[ns]}{n} \sum_{i=1}^{n} Y_i I_{D(x)}(X_i) \right) \right)$$

$n^{-1} \le s \le 1 - n^{-1}, x \in R^d$ and zero otherwise, where $I_{D(x)}(X_i)$ is the indicator of the event

$$[X_i \in D(x)] = [X_i \le x]$$

and

$$\overline{Y I_{D(x)}(X)_n} = n^{-1} \sum_{i=1}^{n} Y_i I_{D(x)}(X_i)$$

βn can also be written as the difference of two means

$$\sqrt{n} w \left(\frac{[ns]}{n} \right) \left(\sum_{i=1}^{[ns]} \frac{Y_i I_{D(x)}(X_i)}{[ns]} - \sum_{i=[ns]+1}^{n} \frac{Y_i I_{D(x)}(X_i)}{(n - [ns])} \right) \tag{4}$$

where

$$w(u) = (u(1-u))^{1/2}.$$

Note that

$$EYI_{[X \in D(x)]} = \int_{D(x)} \phi \, dF.$$

We are considering the partial sums of the response variables Yi aligned according to the Xi-observations. The advantage of βn is that we can use empirical process theory which has much better convergence properties, rather than trying to estimate the function φ directly with kernel or other curve estimation methods which require a large sample size. As will be demonstrated in the simulation study, our approach works well for moderate sample sizes.

In the next section, we state the main results related to the weak convergence of the above process as well as convergence of various related statistics. These results do not assume the stringent condition of bounded variation for the regression function φ and the variance function ψ and strong moment conditions on the error distribution, which were assumed in an earlier paper [5]. Section 3 states the asymptotic results for the weighted bootstrap version of our statistics. This is needed since the asymptotic distributions of our statistics depend on unspecified functions. A new simulation study is conducted in Section 4. Proofs of the results (different from those in [5,6]) are sketched in Section 5.

2. MAIN RESULTS

Assume the following:

(A1) The sequences {Xi} and {εi}, $i = 1, 2, \cdots$ are independent.

(A2) The {Xi} are independent and identically distributed random vectors with distribution function F.

(A3) The functions $\phi, \psi : R^d \to R$, with $\psi \geq 0$, satisfy

$$\int \phi^2 \mathrm{d}F, \int \psi^2 \mathrm{d}F < \infty .$$

(A4) The {εi: 1 < i} are independent and identically distributed with E({εi) = 0 and $E\varepsilon^2 = 1$.

(A5)

$$\int_0^\infty \sqrt{P\left[|\varepsilon| > t\right]} \mathrm{d}t < \infty .$$

Conditions (A1)-(A5) are much weaker than those in [5] and [6] where φ and ψ needed to be of bounded total variation and εi required much higher moment conditions ($\mathrm{E}|\varepsilon_i|^p < \infty$). While the conclusions in [5] and [6] may be stronger (weighted approximations), the requirement of bounded variation for φ and ψ limits their applicability. Here we dispense with this requirement. For example, we permit the case where the Xi have unbounded support and the regression relationship is polynomial (unbounded), provided Xi has finite moments of a suitable degree.

For

$x_i = (x_{i1}, \cdots, x_{id}), i = 1, 2, \cdots$ let

$$x_1 \wedge x_2 = (x_{11} \wedge x_{21}, \cdots, x_{1d} \wedge x_{2d}) .$$

The first main result is:

Theorem 1 Under Assumptions (A1)-(A5) and the null Hypothesis (1),

$$\beta_n \to_D \Gamma, \text{ as } n \to \infty,$$

in the space $D\left(R^d \times [0,1]\right)$, equipped with the uniform norm, where $\Gamma(x,s)$ is mean-zero Gaussian process with covariance function

$$
\begin{aligned}
& E\Gamma(x_1,s_1)\Gamma(x_1,s_1) \\
& =(s_1 \wedge s_2 - s_1 s_2) \\
& \times\left(H_1(x_1 \wedge x_2) - G_1(x_1)G_1(x_2) + H_2(x_1 \wedge x_2)\right),
\end{aligned}
$$

and where

$$
\begin{aligned}
& G_1(x) = \int_{D(x)} \phi \, \mathrm{d}F; \; H_1(x) = \int_{D(x)} \phi^2 \, \mathrm{d}F; \\
& H_2(x) = \int_{D(x)} \psi^2 \mathrm{d}F.
\end{aligned}
$$

Corollary 2 Under the conditions of Theorem 1 and the null Hypothesis (1),

$$
\begin{aligned}
& KS_n \to_D \sup_{x \in R^d} \sup_{0<s<1} \left|\Gamma(x,s)\right|; \\
& MC_n \to_D \int_0^1 \int_{R^d} \left(\Gamma(x,s)\right)^2 \mathrm{d}F \, ds; \\
& KC_n \to_D \sup_{0<s<1} \int_{R^d} \left(\Gamma(x,s)\right)^2 \mathrm{d}F,
\end{aligned} \tag{5}
$$

where

$$
\begin{aligned}
& KS_n = \sup_{x \in R^d} \sup_{0<s<1} \left|\beta_n(x,s)\right|; \\
& MC_n = \int_0^1 \int_{R^d} \left(\beta_n(x,s)\right)^2 \mathrm{d}F \mathrm{d}s; \\
& KC_n = \sup_{0<s<1} \int_{R^d} \left(\beta_n(x,s)\right)^2 \mathrm{d}F.
\end{aligned} \tag{6}
$$

Corollary 2 follows immediately from Theorem 1.

Corollary 3 Under Assumptions (A1)-(A5) and (A3) for φ1 and F1, the test statistics defined by (6) are consistent against the alternative Hypothesis (2).

Under the null Hypothesis (1), $\beta_n(x,s)$ has mean zero for all x and s. The statistic KS_n in Equation (6) is a Kolmogorov-Smirnov type statistic expressing the largest deviation between the two means (before [ns] and after [ns]). Both statistics MC_n and KC_n are Cramérvon Mises type statistics which measures the squaredintegral distance between the two means. The statistic MC_n is the mean (integral over s) such distance, while KC_n is the supremum of the squared-integral distance.

Under the alternative Hypothesis (2), when s < θ, the first part of Expression (4) has mean

$\int_{D(x)} \phi \, dF$ while the term after the minus sign has mean

$\int_{D(x)} \phi^1 \, dF^1$ when s > θ. (see inequality (3)).

3. WEIGHTED BOOTSTRAP APPROXIMATIONS

Since the distribution functions of the Xi and εi, as well as the functional forms of φ and ψ, are not specified, the limiting distributions of all our test statistics will not be known. A weighted bootstrap is proposed here which will approximate these asymptotic distributions.

Let

$$V_1^{(j)}, V_2^{(j)}, \cdots, j = 1, 2, \cdots, m,$$

be m independent sequences of independent and identically distributed random variables, independent of the {εi} and {Xi} sequences. We will consider the processes

$$\beta_n^{(j)}(x,s) = n^{-1/2}\left(\varsigma^{(j)}(x,[ns]) - \frac{[ns]}{n}\varsigma^{(j)}(x,n)\right),$$

for

$n^{-1} \le s \le 1 - n^{-1}, x \in R^d$ zero otherwise, where $j = 1,2,\cdots,m$, and

$$\varsigma^{(j)}(x,k)$$
$$= \left(\sum_{i=1}^{k} Y_i V_i^{(j)} I_{D(x)}(X_i) - \left\{\sum_{\ell=1}^{k} V_\ell^{(j)}\right\}\sum_{i=1}^{k}\frac{Y_i I_{D(x)}(X_i)}{k}\right).$$

Theorem 4 Assume Conditions (A1)-(A5) hold and that (A4)-(A5) holds for the $\{V_i^{(j)}\}$ sequences also. Then, under the null Hypothesis (1), along almost all sequences

$\{(X_i,Y_i), i = 1,2,\cdots\}$ given (Xi,Yi), $i = 1,2,\cdots,n$, $\beta_n^{(j)} \to_D \Gamma$ in the space $D(R^d \times [0,1])$, equipped with the uniform norm, for each j, where Γ is a mean-zero Gaussian process.

Under the conditions of Theorem 4, the same limiting distributions as found in Corollary 2 for statistics based on βn are obtained for each $\beta_n^{(j)}$. Moreover, the statistics based on βn and $\beta_n^{(j)}$, $j = 1,2,\cdots,m$, are asymptotically independent.

It is also important for the bootstrap process to have a limiting distribution under the alternative hypothesis. We obtain:

Theorem 5 If the conditions of Theorem 4 hold for the functions φ1, F1 , then under the alternative Hypothesis (2), along almost all sequences $\{(X_i,Y_i), i = 1,2,\cdots\}$, given (Xi,Yi), $i = 1,2,\cdots,n$, $\beta_n^{(j)} \to_D \Gamma^{11}$ in the space $D\left(R^d \times [0,1]\right)$, equipped with the uniform norm, for each j, where Γ^{11} is a mean-zero Gaussian process.

4. SIMULATIONS

The proposed tests are omnibus tests in that we do not specify the form of the regression function φ. In order to examine how these tests perform, we conducted simulations for two quite different functions, φ(x) = 1 + x + βx2 and φ(x) = 2 sin(βx). The Xi were generated from the uniform U(−3, 3) distribution and the errors εi were standard normals. In each case, 1000 simulations were carried out, each with m = 1000 weighted bootstrap samples. The level of the tests was α = 0.05 throughout. Theorem 3 allows us to choose any mean-zero, variance-one distribution for the bootstrap weights $V_i^{(j)}$. We chose standard normal weights for the simulation since the limiting processes are Gaussian and normal bootstrap weights have been found to work well in simulations for goodness-of-fit types of statistics [7,8]. We use the percentile method. If h(βn) is one of the statistics in Corollary 2, we reject the null hypothesis if $h(\beta_n) > c_n(\alpha)$, where cn(α) is the (1 – α)100-th percentile of the values $h\left(\beta_n^{(j)}\right)$, $j = 1, 2, \cdots, m$. Under the null hypothesis, cn(α)→ c(α), where c(α) is the (1 – α)100-th percentile of $h(\Gamma)$(Theorems 1 and 3). Under the alternative hypothesis, cn(α)→ c11(α), with c11(α), the (1 – α)100-th percentile of $h\left(\Gamma^{11}\right)$and h(βn) →P∞ (Corollary 3 and Theorem 4).

The first power study is for the alternative:

$$H_1 : Y_i = 1 + X_i + 0.4X_i^2 + \psi\varepsilon_i, \quad 1 \le i \le 0.4n$$
$$Y_i = 1 + X_i + \beta X_i^2 + \psi\varepsilon_i, \quad 0.4n < i \le n,$$

where ψ does not depend on Xi and the change-point is at 0.4n. The statistics KSn, KCn and MCn are defined by Equation (6).

The first three rows correspond to the empirical level of the tests under the null hypothesis (β = 0.4) with different choices for ψ, the standard deviation. The breakpoint θ equals 0.4. The empirical level results are reasonably good especially when n > 50, that is they are close to the level of 0.05. The empirical powers of the tests are in the subsequent rows for different β.

As can be seen in Table 1, the power increases as β gets farther from 0.4 (the null Hypothesis). Understandably the power increases in n and decreases in ψ gets larger.

In the second simulation (see Table 2), we consider φ to be a sine function. We conduct our simulations for sample sizes n = 50, 100, 150. The first row of Table 2 is the empirical level of the test (β = 1, no change-point). Again the empirical level is good when n > 50.

We take ψ = 1, but vary the location of the changepoint θn. The power increases as β gets farther from 1 (the null hypothesis). The power increases with the sample size n. Except in a few instances, the power is greatest when the change-point occurs in the middle (θ = 0.5) and decreases as θ is farther from 0.5. The exceptional cases are likely due to simulation error.

Considering that our tests are omnibus in nature, the power results can be seen to be quite good.

5. PROOFS OF THE MAIN RESULTS

Proof of Theorem 1: For $x \in R^d, 0 \le s \le 1$, let

$$\alpha_n(x,s) = \sum_{i=1}^{[ns]} \left(Y_i I_{D(x)}(X_i) - G_1(x) \right) = \alpha_{1n} + \alpha_{2n},$$

where

$$\alpha_{1n}(x,s)=\sum_{i=1}^{[ns]}\left(\phi(X_i)I_{D(x)}(X_i)-G_1(x)\right);$$

$$\alpha_{2n}(x,s)=\sum_{i=1}^{[ns]}\psi(X_i)\varepsilon_i I_{D(x)}(X_i).$$

The collection $\{D(x):x\in R^d\}$ is a Vapnik-Cervonenkis class of sets. Consequently, the collection of indicator functions

$$\left\{I_{D(x)}:x\in R^d\right\}$$

is a Donsker class of functions. For (fixed) φ and ψ,

$$\left\{\phi\, I_{D(x)}:x\in R^d\right\}$$

and

$$\left\{\psi\, I_{D(x)}:x\in R^d\right\}$$

are Donsker classes (Pollard [9]). By Giné and Zinn [10, 11] (see also Corollary 2.9.3 of [12]),

$$n^{-1/2}\left(\alpha_{1n}(x,1),\sum\nolimits_{i=1}^{n}\varepsilon_i(\psi\, I_{D(x)}(X_i)-G_2(x)\right),$$

where

$G_2(x)=\int_{D(x)}\psi\,\mathrm{d}F$ converges in distribution to a vector of tight independent Gaussian processes, η1

and η2, each with zero mean and covariance (j = 1, 2):

$$E\eta_j(x_1)\eta_j(x_2)=\left(H_j(x_1\wedge x_2)-G_j(x_1)G_j(x_2)\right)$$

Consequently,

$$n^{-1/2}\left(\alpha_{1n}(x,1),\alpha_{2n}(x,1)\right)$$

converges in distribution to a vector of tight independent Gaussian processes (η_1,η_2^0) with $E\eta_2^0(x)=0$ and covariance

$$E\eta_2^0(x_1)\eta_2^0(x_2)=H(x_1\wedge x_2).$$

Lastly, (see [12], Theorem 2.12.1)

$$n^{-1/2}\left(\alpha_{1n}(x,s),\alpha_{2n}(x,s)\right)$$

converges in distribution to a vector of tight Gaussian processes $(\eta_1',\eta_2'^0)$ on $D(R^d\times[0,1])$ with zero means and covariances

$$E\eta_1^1(x_1,s_1)\eta_1^1(x_2,s_2)$$
$$=(s_1\wedge s_2)\times\left(H_1(x_1\wedge x_2)-G_1(x_1)G_1(x_2)\right)$$

and

$$E\eta_1^0(x_1,s_1)\eta_1^0(x_2,s_2)=(s_1\wedge s_2)H_1(x_1\wedge x_2).$$

It follows that $n^{-1/2}\alpha_n$ converges in distribution to Γ0 with respect to the uniform metric, where

$$\Gamma^0\equiv n_1^1+\eta_2^0.$$

Since

$$\beta_n(x,s)=n^{-1/2}\left(\alpha_n(x,s)-\frac{[ns]}{n}\alpha_n(x,1)\right),$$

Table 1. Power when $\varphi(x) = 1 + x + 0.4x2$, $\varphi1(x) = 1 + x + \psi x2$ and breakpoint $\theta = 0.4$.

β	ψ	$n = 30$			$n = 50$			$n = 100$		
		KS_n	KC_n	MC_n	KS_n	KC_n	MC_n	KS_n	KC_n	MC_n
0.4	1	0.038	0.046	0.051	0.044	0.043	0.051	0.057	0.050	0.054
	2	0.040	0.047	0.053	0.047	0.060	0.063	0.052	0.049	0.048
	3	0.043	0.040	0.046	0.056	0.048	0.050	0.056	0.049	0.048
0.2	1	0.096	0.142	0.175	0.119	0.228	0.252	0.222	0.408	0.427
	2	0.086	0.057	0.076	0.088	0.088	0.097	0.121	0.136	0.139
	3	0.063	0.059	0.071	0.090	0.087	0.088	0.112	0.093	0.098
0.0	1	0.264	0.309	0.387	0.448	0.596	0.644	0.708	0.916	0.933
	2	0.159	0.120	0.163	0.257	0.261	0.293	0.454	0.488	0.510
	3	0.104	0.068	0.092	0.169	0.154	0.166	0.287	0.266	0.280
-0.2	1	0.465	0.383	0.517	0.748	0.768	0.807	0.974	0.986	0.988
	2	0.297	0.197	0.275	0.481	0.438	0.470	0.802	0.778	0.784
	3	0.202	0.137	0.167	0.300	0.241	0.269	0.539	0.485	0.502
-0.4	1	0.663	0.443	0.591	0.916	0.826	0.888	1.000	0.998	0.999
	2	0.474	0.301	0.396	0.698	0.549	0.614	0.950	0.920	0.925
	3	0.292	0.175	0.219	0.476	0.384	0.415	0.801	0.697	0.722

Table 2. Power when φ(x) = 2 sin(x), φ1(x) = 2 sin(βx), and ψ = 1 with different breakpoints θ.

		$n = 50$			$n = 100$			$n = 150$		
β	θ	KS_n	KC_n	MC_n	KS_n	KC_n	MC_n	KS_n	KC_n	MC_n
1.0		0.052	0.051	0.048	0.046	0.046	0.056	0.051	0.045	0.051
1.2	0.3	0.050	0.053	0.051	0.068	0.077	0.076	0.078	0.079	0.087
	0.4	0.054	0.063	0.060	0.075	0.087	0.091	0.079	0.089	0.096
	0.5	0.057	0.059	0.061	0.075	0.090	0.094	0.081	0.093	0.102
	0.6	0.058	0.055	0.054	0.065	0.073	0.071	0.086	0.092	0.098
	0.7	0.053	0.054	0.058	0.069	0.073	0.080	0.069	0.066	0.067
1.4	0.3	0.075	0.104	0.111	0.101	0.182	0.193	0.127	0.254	0.263
	0.4	0.088	0.127	0.131	0.130	0.233	0.246	0.188	0.344	0.349
	0.5	0.090	0.124	0.131	0.136	0.252	0.255	0.198	0.359	0.361
	0.6	0.092	0.131	0.139	0.140	0.246	0.248	0.203	0.368	0.359
	0.7	0.062	0.098	0.105	0.108	0.189	0.189	0.139	0.278	0.277
1.6	0.3	0.109	0.148	0.170	0.174	0.331	0.345	0.259	0.489	0.476
	0.4	0.121	0.190	0.215	0.243	0.405	0.425	0.382	0.597	0.598
	0.5	0.125	0.219	0.222	0.263	0.448	0.455	0.416	0.640	0.638
	0.6	0.115	0.193	0.197	0.254	0.423	0.429	0.383	0.603	0.592
	0.7	0.092	0.150	0.159	0.188	0.349	0.340	0.285	0.495	0.484
1.8	0.3	0.118	0.206	0.208	0.235	0.443	0.464	0.425	0.625	0.626
	0.4	0.166	0.262	0.292	0.372	0.549	0.558	0.585	0.745	0.742
	0.5	0.169	0.280	0.295	0.396	0.592	0.595	0.636	0.760	0.766
	0.6	0.148	0.256	0.260	0.380	0.580	0.583	0.588	0.733	0.729
	0.7	0.111	0.207	0.200	0.263	0.461	0.459	0.421	0.648	0.632
2.0	0.3	0.127	0.225	0.238	0.297	0.484	0.488	0.527	0.695	0.679
	0.4	0.181	0.295	0.307	0.437	0.626	0.628	0.692	0.806	0.810
	0.5	0.204	0.319	0.327	0.506	0.659	0.661	0.765	0.822	0.815
	0.6	0.190	0.284	0.319	0.446	0.622	0.620	0.705	0.802	0.790
	0.7	0.143	0.223	0.234	0.312	0.497	0.495	0.514	0.699	0.674

We obtain βn→D Γ, as n→∞, where

$$\Gamma =_D \Gamma^0(x,s) - s\Gamma^0(x,1)$$

and has covariance given in the statement of Theorem 1. Hence Theorem 1 is proven.

By viewing α2n(x, [ns]) as a strong martingale and using the functional central limit theorem for strong martingales by Ivanoff [13], one can drop the assumption (A5) and get convergence in the Shorohod topology. (see [14]).

Proof of Corollary 3: Assume the alternative Hypothesis (2). Then,

$$\beta_n(x,s) = n^{-1/2}\left(\alpha_n^{11}(x,s) - \frac{[ns]}{n}\alpha_n^{11}(x,1)\right)$$
$$+\gamma_n(x,s),$$

where, for s < θ,

$$\gamma_n(x,s) = \sqrt{n}\left(\frac{[ns]}{n}\right)\left(\frac{n-[n\theta]}{n}\right)\left(G_1(x) - G_1^1(x)\right)$$
$$\alpha_n^{11}(x,s) = \alpha_n(x,s),$$

and for $s > \theta$,

$$\alpha_n^{11}(x,s) = \alpha_n(x,\theta) + \alpha_n^1(x,s) - \alpha_n^1(x,\theta),$$
$$\gamma_n(x,s) = \sqrt{n}\left(\frac{[n\theta]}{n}\right)\left(\frac{n-[ns]}{n}\right)\left(G_1(x) - G_1^1(x)\right),$$

with

$$\alpha_n^1(x,s) = \sum_{i=1}^{[ns]}\left(Y_i I_{D(x)}(X_i) - G_1^1(x)\right),$$

and

$$G_1^1(x)=\int_{D(x)}\phi^1 \mathrm{d}F^1$$

.

If Condition (A3) holds for the functions φ1 and F1, then under the alternative Hypothesis (2), $n^{-1/2}\alpha_n^{11}\to_D\Gamma^{11}$ in the space $D(R^d\times[0,1])$ with the uniform metric, where Γ11 is a zero-mean Gaussian process. Strong consistency of the statistics nKS, nKC, nMC follows. (See the proof of Theorem 4.1 in [5]).

Proof of Theorem 4: Let

$$Z_i(x)=Y_i I_{D(x)}(X_i)$$

.

By the multivariate CLT, along almost all sequences,

$$\{(X_i,\varepsilon_i),i=1,2,\cdots\},$$

$$\xi_n(x,s)\equiv n^{-1/2}\sum_{i=1}^{[ns]}V_i Z_i(x)$$

is a sum of iid random variables with mean zero and a resulting covariance matrix having entries of the form

$$n^{-1}\sum_{i=1}^{\lfloor(s_1\wedge s_2)\rfloor}Y_i^2 I_{D(x_1\wedge x_2)}(X_i),$$

which converge, uniformly in (x1, s1) and (x2, s2),

$$(s_1\wedge s_2)(H_1(x_1\wedge x_2)+H_2(x_1\wedge x_2))$$

.

Similarly, the corresponding conditional covariance of

$$\zeta^{(j)}(x,[ns])$$

converges uniformly to

$(s_1 \wedge s_2)(H_1(x_1 \wedge x_2) - G_1(x_1 \wedge x_2) + H_2(x_1 \wedge x_2))$ and the corresponding conditional covariance of $\beta^{(j)}(x,[ns])$ converges uniformly to that of Γ.

By the multiplier inequalities ([12], Lemma 3.6.7),

$$E_V \left\| n^{-1/2} \sum_{i=1}^{n} V_i Z_i \right\|_{\mathfrak{I}}^{*}$$

$$\leq 2(n_0 - 1) n^{-3/2} \sum_{i=1}^{n} \|Z_i\|_{\mathfrak{I}}^{*} E_V \left(\max_{1 \leq i \leq n} |V_i| \right)$$

$$+ 2\int_0^{\infty} \sqrt{P(V_1 > t)}\,dt \max_{n_0 \leq k n} E_R \left\| k^{-1/2} \sum_{i-n_0}^{n} Z_{R_i} \right\|_{\mathfrak{I}}^{*},$$

where $(R_1, \cdots, R_n)$ is uniformly distributed on the set of all permutations of $\{1, 2, \cdots, n\}$ and is independent of $(Z_1, \cdots, Z_n)$. In our case, $\mathfrak{I}$ is the collection of functions

$$\{I_{D(x)} : x \in R^d\}.$$

also,

$$E.\| \cdot \|_{\mathfrak{I}}^{*}$$

denotes (uniform) measurable covers with respect to all variables jointly. Using Theorem 1, as in the proof of Theorem 3.6.13 in [12], x n(x, 1) is Donsker. It follows that ([12], Theorem 2.12.2) xn(x, s) is Donsker, and Theorem 4 for b(j) follows.

Proof of Theorem 5: The proof is similar to that of Theorem 4 but where the limiting covariance of is that of G11 instead of Γ.

6. CONCLUDING REMARKS

In this paper, we have constructed test statistics using $F = \left\{ I_{D(x)} : x \in R^d \right\}$ the lower orthants. It is clear (see [12]) that other Donsker classes of functions can be used. Examples are: the collection of the closed balls, or the collection of half-spaces. The method of proof in [5,6,15] could not handle such classes. One can also obtained weak convergence of the weighted version of the process where the weight functions satisfy the Chibisov-O'Reilly conditions (see [5,6], or [16], p. 462).

The simulations show that our tests have good power considering the omnibus nature of the procedure. By avoiding curve estimation (e.g. using kernel methods to estimate f), we are able to provide practical tests for moderate sample sizes.

REFERENCES

1. C. Lee,M. Csörgő and L. Horváth, "Limit Theorems in Change-Point Analysis," Wiley, Chichester, 1997.

2. M. Csörgő, L. Horváth and B. Szyszkowicz, "Integral Tests for Suprema of Kiefer Processes with Application," Statist, Decisions, Vol. 15, No. 4, 1997, pp. 365-377.

3. M. Csörgő and B. Szyszkowicz, "Weighted Multivariate Empirical Processes and Contiguous Change-Point Analysis," Change-Point

Problems, IMS Lecture Notes Monograph Series, Vol. 23, 1994, pp. 93-98.

4. M. Csörgő and B. Szyszkowicz, "Applications of MultiTime Parameter Processes to Change-Point Analysis, In: M. Fisz, Ed., Probability Theory and Mathematical Statistics, John Wiley & Sons, Inc., New York, 1994, pp. 159-222.

5. M. D. Burke, "Limit Theorems for Change-Point Detection in Nonparametric Regression Models," Acta Mathematica (Szeged), Vol. 73, No. 3-4, 2007, pp. 865-882.

6. M. D. Burke, "Approximations for a Multivariate Hybrid Process with Applications to Change-Point Detection," Mathematical Methods of Statistics, Vol. 19, No. 2, 2010, pp. 121-135.

7. M. D. Burke, "A Gaussian Bootstrap Approach to Estimation and Tests," In: B. Szyszkowicz, Ed., Asymptotic Methods in Probability and Statistics, Elsevier, Amsterdam, 1998, pp. 697-706.

8. M. D. Burke, "Multivariate Tests-of-Fit and Uniform Confidence Bands Using a Weighted Bootstrap," Statistics and Probability Letters, Vol. 46, No. 1, 2000, pp. 13-20.

9. D. Pollard, "A Central Limit Theorem for Empirical Processes," Journal of the Australian Mathematical Society A , Vol. 33, No. 2, 1982, pp. 235-248.

10. E. Giné and J. Zinn, "Bootsrapping General Empirical Measures," Annals of Probability, Vol. 18, No. 2, 1984, pp. 851-869. doi:10.1214/aop/1176990862

11. E. Giné and J. Zinn, "Lectures on the Central Limit Theorem for Empirical Processes," Lecture Notes in Mathematics, Vol. 1221, 1986, pp. 50-113.

12. A. W. Van der Vaart and J. A. Wellner, “Weak Convergence and Empirical Processes: With Applications to Statistics,” Springer, New York, 1996.

13. A. G. Ivanoff, “Stopping Times and Tightness for Multiparameter martingales,” Statistics & Probability Letters, Vol. 28, No. 2, 1996, pp. 111-114.

14. M. D. Burke, “Testing Regression Models: A Strong Martingale Approach,” Fields Institute Communications. 2004.

15. L. Horváth, “Approximations for hybrids of empirical and partial sums processes,” Journal of Statistical Planning and Inference, Vol. 88, No. 1, 2000, pp. 1-18.

16. G. R. Shorack and J. A. Wellner, “Empirical Processes with Applications to Statistics,” Wiley, New York, 1986.

Chapter 9

Causal inference, probability theory, and graphical insights

Stuart G. Baker*

*Biometry Research Group, National Cancer Institute, Bethesda, MD 20892 USA.

ABSTRACT

Causal inference from observational studies is a fundamental topic in biostatistics. The causal graph literature typically views probability theory as insufficient to express causal concepts in observational studies. In contrast, the view here is that probability theory is a desirable and sufficient basis for many topics in causal inference for the following two reasons. First probability theory is generally more flexible than causal graphs: besides explaining such causal graph topics as M-bias (adjusting for a collider) and bias amplification and attenuation (when adjusting for instrumental variable), probability theory is also the foundation of the paired availability design for historical controls, which does not fit into a causal graph framework. Second probability theory is the basis for insightful graphical displays including the BK-Plot for understanding Simpson's paradox with a binary confounder, the BK2-plot for understanding bias amplification and attenuation in the presence of an unobserved binary confounder, and the PAD-Plot for understanding the principal stratification component of the paired availability design.

KEYWORDS

BK-Plot, causal graph, confounder, instrumental variable, observational study, Simpson's paradox

1. INTRODUCTION

Perhaps the greatest concern in the analysis of observational data in a clinical setting is the lack of complete knowledge as to why a person receives a particular treatment [1]. For this reason the causal analysis of observational studies is a challenge. Two common types of observational studies are multivariate adjustments with concurrent controls and analyses based on before-and-after studies.

Causal graphs are one approach to understanding bias with multivariate adjustment when estimating treatment effect in observational studies [2]. In these multivariate adjustments lack of knowledge as to why a person receives treatment can be viewed as an unobserved confounder that directly influences treatment received and outcome. An unobserved confounder leads to biased estimates of causal effect if not included in a multivariate adjustment. Examples of unobserved confounders are unrecorded information on treatment history, disease history, or symptoms.

The causal graph literature has contributed new ideas to understanding bias in multivariate adjustments, such as M-bias when incorrectly adjusting for a collider [3–10] and bias amplification or attenuation when incorrectly adjusting for an instrumental variable in the presence of an unobserved confounder [11–16]. Some of the causal graph literature claims that probability theory is insufficient for understanding causal inference [2]. The view here is that probability theory is a desirable and sufficient basis for causal inference in multivariate adjustments if there is no adjustment for a variable that is a consequence of treatment. Adjusting for a variable that both affects outcome and is a consequence of treatment is well-known to yield biased estimates of treatment effect [17].

A probability theory viewpoint to understanding casual inference in observational studies has two desirable aspects. First probability theory is flexible: besides explaining M-bias and bias amplification and attenuation, probability theory is the basis for the paired availability design for historical controls [18, 19], a method outside the purview of casual graphs. Second probability theory leads to graphical insights: the BK-Plot for understanding Simpson's paradox with a binary confounder, the BK2-plot for understanding bias amplification and attenuation in the presence of an unobserved binary confounders, and the PAD-Plot for understanding the principal stratification [20] component of the paired availability design. Graphical approaches have a long history of providing insight in mathematics such as the "look-see" proof of the Pythagorean theorem [21] and the sum of an infinite geometric series [22].

To put both probability theory and causal graphs into perspective, it is worth noting that there are topics in causal inference for biology and medicine in which neither applies. Examples include downward causation [23] and simply distinguishing cause from effect as whether mutations cause cancer or cancer causes mutations [24–26].

2. SIMPSON'S PARADOX: THE BK-PLOT

Simpson's paradox is perhaps best summarized by the iconic phrase of the noted statistician Thomas Louis "Good for men, good for women, bad for people" [27], which applies to the effect of treatment on outcome in causal graph *(a)* where *X* is treatment, *Y* is outcome, and *U* is sex. The correct approach is to adjust for *U* yielding the conclusion that treatment is beneficial.

There are two major types of discussions regarding Simpsons' paradox. In both cases (essentially by definition) if the paradox is examined in the appropriate way, the paradox disappears. One type of discussion focuses why adjustment for *U* is correct in causal graph *(a)* while an unadjusted estimate is correct for causal graphs *(b)* and *(c)* [2, 28, 29]. In causal graph *(b) U* is a consequence of treatment, so adjustment is not appropriate. Causal graph *(c)* illustrates M-bias, a topic of considerable debate in *Statistics in Medicine* [4–9] The key aspect of this graph is that *U* is a collider (two arrows point to it). Because *U* is a collider, *X* and *Y* are independent on the back-door path ($X \leftarrow Q \rightarrow U \leftarrow R \rightarrow Y$) but adjusting for *U* makes *X* and *Y* dependent on this back-door path. Therefore for causal graph *(c)* adjustment on *U* is not appropriate Although this is a well-known result in the causal graph literature [2], a simple proof based solely on probability theory is given in the Appendix A. In more complicated causal graphs involving multiple colliders, the back-door criterion can be used to determine if *X* and *Y* are independent on the back-door path [2]. An open question is how often M-bias arises in practice, as an analyst would need to adjust for *U* but not *R* or *Q*. However, if *Q* were known, the analyst would adjust for *Q* so M-bias would not arise.

A second type of discussion of Simpson's paradox involves understanding why a reversal of signs occurs with the crude versus the adjusted risk difference for the scenario depicted in causal graph *(a)* when *U* is a binary variable. Table 1 presents a numerical example. Here the BK-Plot provides useful insight by graphically summarizing the probability calculations with mixtures of binary variables. The BK-Plot was developed independently by Jeon *et al.* [30] and Baker and Kramer [27] and given its name by Howard Wainer [31]. The BK-Plot has also been used to illustrate calculations involving missing data [32], the transitive fallacy in randomized trials [33], and binary surrogate endpoints [34].

Table 1. Hypothetical data illustrating bias when *U* is a confounder. The conditional *RD* is 0.10. The crude *RD* is −0.20.

	confounder								
	U=0			U=1			U=?		
	outcome			outcome			outcome		
treatment	Y=0	Y=1	pr(Y=1)	Y=0	Y=1	pr(Y=1)	Y=0	Y=1	pr(Y=1)
X=0	80	20	0.20	60	240	0.80	140	260	0.65
X=1	210	90	0.30	10	90	0.90	220	180	0.45
RD			0.10			0.10			−0.20

The arrows in causal graph *(a)* imply a joint probability distribution (combining both front and back-door paths) that is initially factored as pr(*Y*=1, *U*=*u*, *X* =*x*) = pr(*Y*=*1* | *U*=*u*, *X*=*x*) pr(*X*=*x* | *U*=*u*) pr(*U*=*u*). The first step in formulating a BK-Plot is to rewrite this joint distribution as

$$\mathrm{pr}(Y=1, U=u, X=x)=\mathrm{pr}(Y=1|X=x)\mathrm{pr}(U=u|X=x)\mathrm{pr}(X=x). \tag{1}$$

In this formulation the causal effect is summarized by a single parameter Δ so that

$$\mathrm{pr}(Y=1|U=u, X=1)=\mathrm{pr}(Y=1|U=u, X=0)+\Delta. \tag{2}$$

The conditional risk difference is $\mathrm{pr}(Y=1|\ U=u, X=1) - \mathrm{pr}(Y=1|\ U=u, X=0) = \Delta$. The crude risk difference, which is biased because it does not adjust for *U*, is

$$\text{crude}RD = \text{pr}(Y=1|X=1) - \text{pr}(Y=1|X=0), \text{ where } \text{pr}(Y=1|X=x) = \Sigma_u \text{pr}(Y=1|U=u, X=1)\text{pr}(U=u|X=x). \quad (3)$$

The adjusted risk difference equals the crude risk difference with pr(*U*=*u*) substituted for pr(*U*=*u* | *X*=*x*) in equation (3),

$$\text{adjusted}RD = \Sigma_u \text{pr}(Y=1|U=u, X=1)\text{pr}(U=u|X=1) - \Sigma_u \text{pr}(Y=1|U=u, X=0)\text{pr}(U=u|X=0). \quad (4)$$

The BK-Plots at the top of Figure 2 displays the above algebraic equations corresponding to Table 1. Each diagonal line plots pr(*Y*=1| *x*) as a function of pr(*U*=1 | *x*). The diagonal lines are parallel because there is no interaction between treatment *X* and confounder *U* in equation (2). The conditional *RD* is 0.30–0.20 = 0.10 for *U*=0 and 0.90–0.80 = 0.10 for *U*=1.

The plot on the top left of Figure 2 graphically shows the computation of the crude *RD* using different dashed vertical lines for pr(*U*=1 | *X*=0) and pr(*U*=1 | *X*=1), where the crude *RD* is the difference between the horizontal dashed lines. Here the crude *RD* is {(0.30) (0.75) + (0.90) (0.25)} – {(0.20) (0.25) +(0.80) (0.75)} =0.45 – 0.65 = –0.20.

The plot on the top right of Figure 2 shows the computation of the adjusted *RD* using a single dashed vertical line for pr(*U*=1 | *x*) = pr(*U*=1). Graphically the adjusted risk difference is the vertical difference between the diagonal lines. Here the adjusted *RD* is {0.30 pr(*U*=0) + 0.90 pr(*U*=1)} – {0.80 pr(*U*=1) + 0.20 pr(*U*=0)}= 0.10.

The BK-Plot visually shows that the crude *RD* is biased when the distribution of *U* differs by treatment group, as indicated by the horizontal shift in the dashed vertical lines that leads to a reversal of signs for the crude *RD* versus the adjusted *RD*. If there were an interaction between *U* and *X* in equation (2), the diagonal lines would not be parallel, but the reversal of signs could still occur.

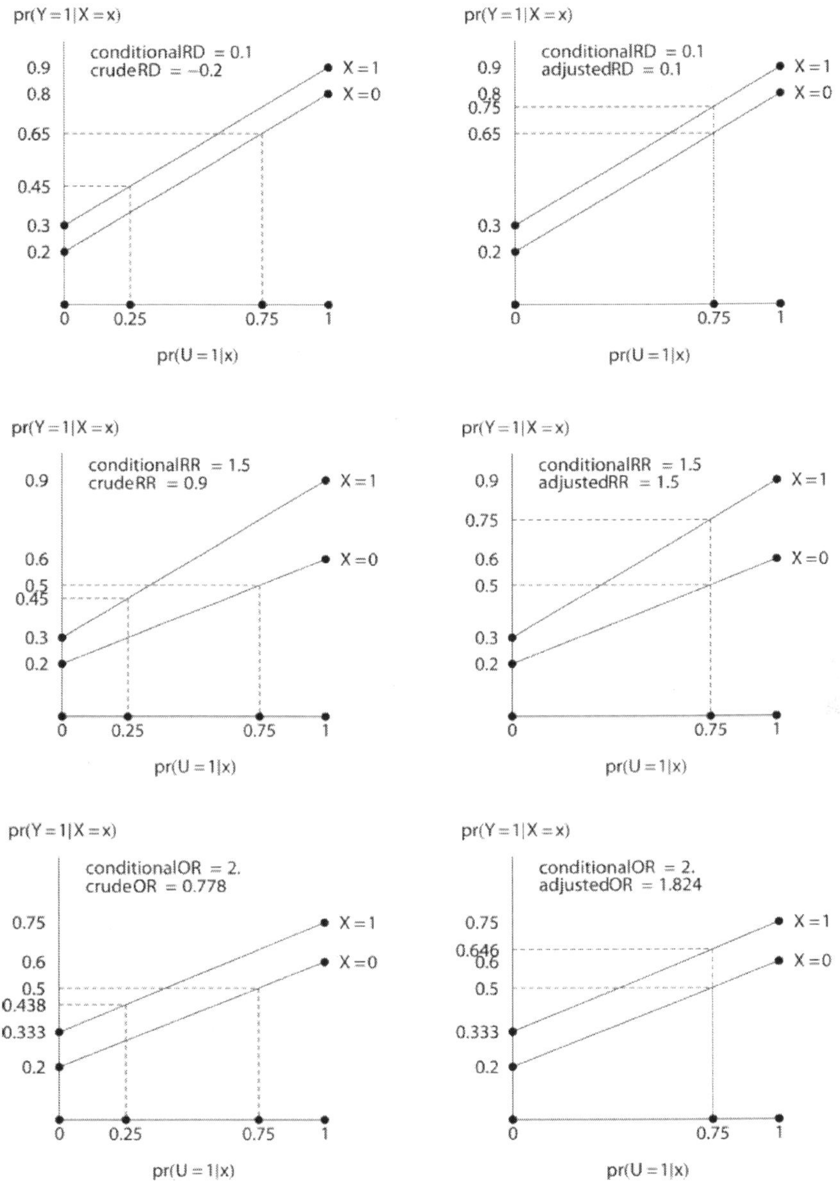

Figure 2. BK-Plots involving risk difference, relative risk, and odds ratio.

3. AN UNOBSERVED BINARY CONFOUNDER: THE BK-PLOT

The BK-Plot also provides insight into bias from an unobserved binary confounder. Examples of unobserved binary confounders include presence or absence of previous treatments [1], complete or incomplete removal of tumors prior to therapy [1], presence or absence of intense pain in women in labor [18], or early or late treatment initiation relative to the development of disease. An unobserved binary confounder also summarizes a combination of factors for low or high risk of outcome that is related to choice of treatment.

Figure 2 can be viewed as BK-Plot for an unobserved binary confounder under three scales: *RD*, relative risk (*RR*) and the odds ratio (*OR*) [35]. It is well known that if there is no interaction between *X* and *U*, the conditional *RD* equals the adjusted *RD*, the conditional *RR* equals the adjusted *RR*, but the conditional *OR* does not equal the adjusted *OR* [35,36]. Consequently *RD* and *RR*, but not *OR*, are said to be collapsible [37]. Collapsibility of *RD* and *RR*, but not *OR*, is illustrated on the right side of Figure 2. In this scenario with no interaction between *X* and *U*, adjusted values of *RD* and *RR*, but not *OR*, are valid for other distributions of *U*, an important consideration in the meta-analysis of randomized trials when *U* is unobserved [35].

4. BIAS AMPLIFICATION AND ATTENUATION: BK-2 PLOT

In recent years there has been considerable interest in understanding how bias from an unobserved confounder is amplified or attenuated by adjustment for an instrumental variable, which is a variable that only influences outcome through treatment [11–16]. One example of an instrumental variable is the clinic where a patient is seen, when each clinic preferentially administers a given treatment but otherwise provides the same patient management. Another example is glaucoma diagnosis, which affects whether statin or glaucoma drugs are received, but is not thought to be directly to incidence of hip fracture [26]. In practice an investigator may be uncertain if a variable is instrumental, so includes it in the multivariate adjustment in case it may be a confounder. An important issue is how this incorrect adjustment affects bias.

Two motivating examples are presented in which the statistic of interest is *RD*. Variable *U* is the confounder and *Z* is the instrumental variable. The hypothetical data in Table 2 illustrate bias amplification. The conditional *RD* (for each stratum of *U* and *Z*) is 0.10. The crude *RD* (combining tables for *U* and *Z*) is −0.068, which translates into a bias of magnitude 0.168. In contrast the *Z*-adjusted *RD* (after combining tables over *U*) is −0.11 which translates into a larger bias of magnitude 0.21—hence bias amplification. The hypothetical data in Table 3 illustrate bias attenuation. The conditional *RD*, which is the causal effect, is 0.10. The crude *RD* is −0.25, which translates into a bias of magnitude 0.35. In contrast the *Z*-adjusted *RD* is −0.11, which translates into a smaller bias of magnitude 0.21 –hence bias attenuation. The BK2-Plots in Figures 3 and and44 visually explain these results.

We discuss bias amplification and attenuation in the situations depicted in the causal graphs (*d*) and (*e*) in Figure 1 where *U* is an unknown binary confounder. These causal graphs summarize a key assumption, namely *Y* is independent of *Z* given *X* and *U*. Causal graph (*d*) defines *Z* as an instrumental variable [27]. For causal graph (*d*) the joint distribution of the variables, as indicated by the arrows, is $pr(Y=1, U=u, Z=z, X=x) = pr(Y=1 \mid x, u)\, pr(X=x \mid u, z)\, pr(Z=1)\, pr(U=u)$. For causal graph (*e*), the joint distribution of the variables, as indicated by the arrows, is $pr(Y=1, U=u, Z=z, X=x) = pr(Y=1 \mid x, u)\, pr(X=x \mid u, z)\, pr(Z=1, U=u)$. We can rewrite the joint distribution of the variables in causal graphs *(d)* and *(e)* as

$$pr(Y=1,U=u,Z=z,X=x)=pr(Y=1|x,u)pr(U=u|x,z)pr(Z=1|x)pr(X=x) \quad (5)$$

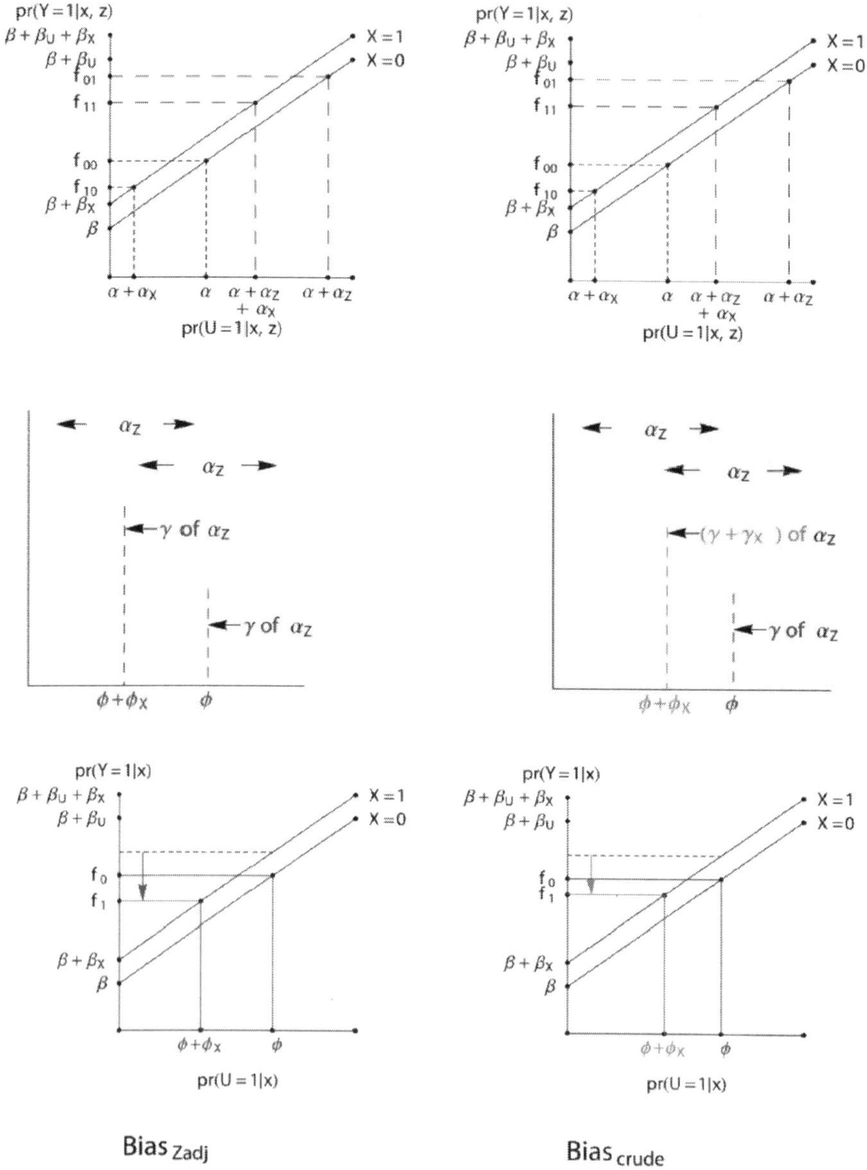

Figure 3. BK2-Plot showing bias amplification.

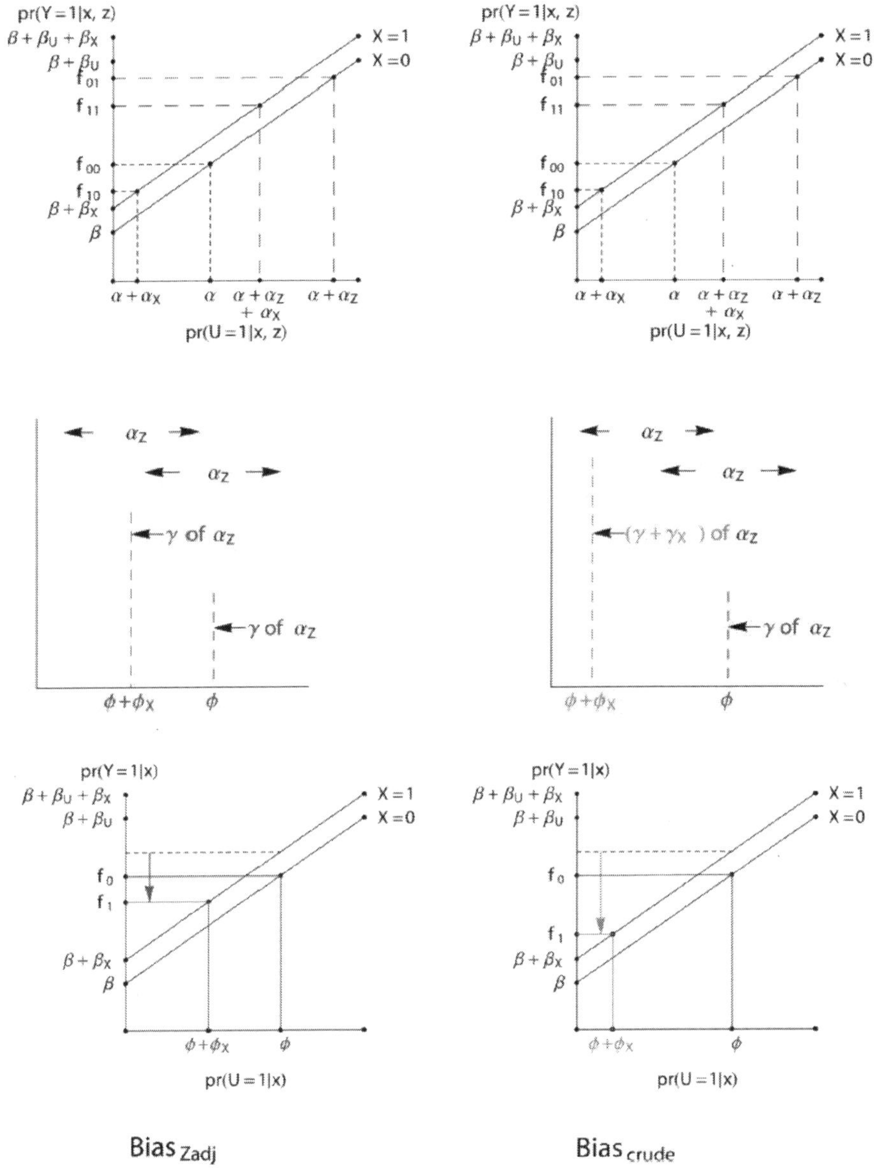

Figure 4. BK2-Plot showing bias attenuation.

Table 2. Hypothetical data illustrating bias amplification. The conditional *RD* is 0.10. The crude *RD* is −0.068, which translates into a bias of magnitude 0.168. The Z-adjusted *RD* is −0.11, which translates into a larger bias of magnitude 0.21.

		confounder								
		U=0			*U*=1			*U*=?		
		outcome			outcome			outcome		
clinic	treatment	*Y*=0	*Y*=1	pr(*Y*=1)	*Y*=0	*Y*=1	pr(*Y*=1)	*Y*=0	*Y*=1	pr(*Y*=1)
Z=0	*X*=0	240	60	0.20	20	180	0.90	260	240	0.48
	X=1	239	103	0.30	0	38	1.00	239	141	0.37
	RD			**0.10**			**0.10**			**−0.11**
Z=1	*X*=0	40	10	0.20	45	405	0.90	85	415	0.83
	X=1	174	74	0.30	0	372	1.00	174	446	0.72
	RD			**0.10**			**0.10**			**−0.11**
Z=?	*X*=0							345	655	0.655
	X=1							413	587	0.587
	RD									**−0.068**

Because the focus is the effect of *X* on *Y*, it is not a concern that equation (5) does not preserve the information in causal graph (*d*) that *U* and *Z* are independent.

To gain insight into how bias amplification and attenuation plays out, a particular parameterization is needed. The BK2-Plot is based on the following parameterization for equation (5),

- Model Y: $\text{pr}(Y=1 \mid x, u) = \beta + \beta_U u + \beta_X x$,
- Model U: $\text{pr}(U=1 \mid x, z) = \alpha + \alpha_Z z + \alpha_X x + \alpha_{ZX} x z$,
- Model Z: $\text{pr}(Z=1 \mid x) = \gamma + \gamma_X x$,

where the binary variables *u, x,* and *z* are set equal either 0 or 1. In Model Y, the causal effect of *X* on *Y* is denoted by β_X. For simplicity, there is no modification of the causal effect by an interaction between *X* and U. The BK2-Plot implicitly requires that the parameters yield probabilities between 0 and 1. Because the purpose of the BK2-Plot is explanation, and not estimation, there is no concern about parameters lying outside admissible values.

The BK2-Plot involves the following two algebraic derivations of risk difference, which extend those for a single binary confounder [13, 14, 28] to include the additional variable Z

Table 3. Hypothetical data illustrating bias amplification. The conditional RD, which is the causal effect, is 0.10. The crude *RD* is −0.25, which translates into a bias of magnitude 0.35. The Z-adjusted *RD* is −0.11, which translates into a smaller

		confounder								
		U=0			*U*=1			*U*=?		
		outcome			outcome			outcome		
clinic	treatment	*Y*=0	*Y*=1	pr(*Y*=1)	*Y*=0	*Y*=1	pr(*Y*=1)	*Y*=0	*Y*=1	pr(*Y*=1)
Z=0	*X*=*0*	240	60	0.20	20	180	0.90	260	240	0.48
	X=*1*	567	243	0.30	0	90	1.00	567	333	0.37
	RD			**0.10**			**0.10**			**−0.11**
Z=1	*X*=*0*	40	10	0.20	45	405	90	85	415	0.83
	X=*1*	28	12	0.30	0	60	1.00	28	72	0.72
	RD			**0.10**			**0.10**			**−0.11**
Z=?	*X*=*0*							345	655	0.655
	X=*1*							595	405	0.405
	RD									−0.25

4.1 Derivation I

Let

$$f_{xz} = \text{pr}(Y=1 \mid x, z) = \Sigma_u \, \text{pr}(Y=1 \mid x, z, u) \, \text{pr}(U=u \mid x, z)$$

so that, under the models,

$$f_{00}=\beta(1-\alpha)+(\beta+\beta_U)\alpha, \quad (6)$$

$$f_{10}=(\beta+\beta_X)(1-\alpha-\alpha_X)+(\beta+\beta_X+\beta_U)(\alpha+\alpha_X), \quad (7)$$

$$f_{01}=\beta(1-\alpha-\alpha_Z)+(\beta+\beta_U)(\alpha+\alpha_Z), \quad (8)$$

$$f_{11}=(\beta+\beta_X)(1-\alpha-\alpha_Z-\alpha_X-\alpha_{ZX})+(\beta+\beta_X+\beta_U)(\alpha+\alpha_Z+\alpha_X+\alpha_{ZX}). \quad (9)$$

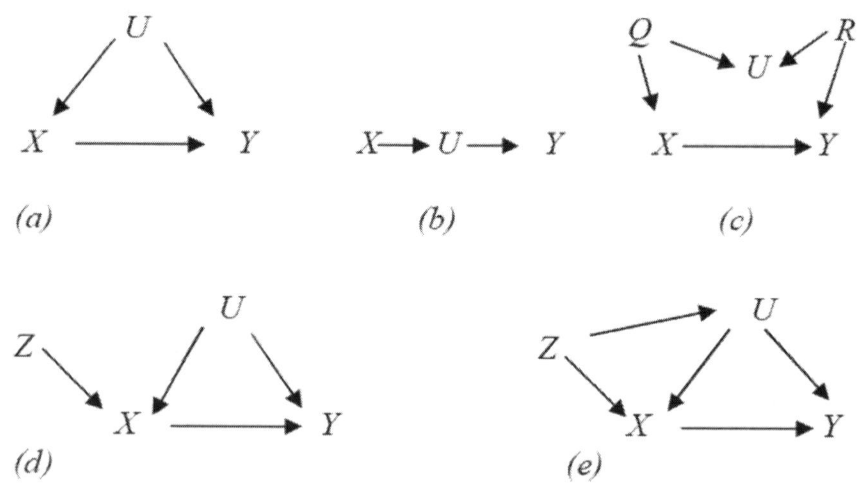

Figure 1. Causal graphs where *X* is treatment, *U* is a confounder, and *Y* is outcome.

Let $f_x = \mathrm{pr}(Y=1 \mid X=x)$. The crude risk difference is the risk difference not adjusting for any variable, namely,

$$RD_{crude}=f_1 - f_0.=\Sigma_z f_{1z}\text{pr}(Z=z|X=1) - \Sigma_z f_{0z}\text{pr}(Z=z|X=0)$$

(10)

$$=\beta_X+\beta_U\alpha_X+\beta_U\gamma_X\alpha_Z+\beta_U(\gamma+\gamma_X)\alpha_{ZX}. \quad (11)$$

The risk difference conditional on stratum *z* is

$$RD_{cond(z)}=f_{1z} - f_{0z}=\beta_X+\beta_U\alpha_X+\beta_U\alpha_{ZX}z. \quad (12)$$

The *Z*-adjusted risk difference is obtained by substituting pr(*Z*=*z*) = γ for pr(*Z*=*z* | *X*=*x*) in equation (10) to yield

$$RD_{Zadj}=\Sigma_z f_{1z}\gamma - \Sigma_z f_{0z}\gamma=\beta_X+\beta_U\alpha_X+\beta_U\gamma\alpha_{ZX}.$$

(13)

As required, $RD_{Zadj} = RD_{crude}$ when $\gamma_X = 0$.

4.2 Derivation II

Using the identity pr(*U*=1|*x*)= Σ_z pr(*U*=1 | *x, z*) pr(*Z*=*z* | *x*), let

$$\phi=\text{pr}(U=1|X=0)=\alpha(1 - \gamma)+(\alpha+\alpha_Z)\gamma, \quad (14)$$

$$\phi+\phi_X=\text{pr}(U=1|X=1)=(\alpha+\alpha_X)(1-\gamma-\gamma_X)+(\alpha+\alpha_Z+\alpha_X+\alpha_{ZX})(\gamma+\gamma_X),$$

(15)

where $\varphi_X = \alpha_X + \gamma_X\alpha_Z + (\gamma+\gamma_X)\alpha_{ZX}$. The identity $f_x = \Sigma_u$ pr(*Y*=1 | *x, u*) pr(*U*=*u* | *x*) gives

$$f_0=\beta(1-\phi)+(\beta+\beta_U)\phi=\beta+\beta_U\phi,$$
$$f_1=(\beta+\beta_X)(1-\phi-\phi_X)+(\beta+\beta_U+\beta_X)(\phi+\phi_X)=\beta+\beta_X+\beta_U\phi+\beta_U\phi_X.$$

(16)

As will be shown, the BK2-Plot graphically derives RD_{crude} from equation (16) and graphically derives RD_{Zadj} by setting $\gamma_X=0$ in equation (15) before substitution into equation (16).

4.3 Bias amplification and attenuation

Bias amplification (attenuation) refers to a larger (smaller) bias when estimating causal effect in the presence of the unobserved confounder when using RD_{Zadj} instead of RD_{crude}. These biases are

$$Bias_{crude}=RD_{crude}-\beta_X=\beta_U\alpha_X+\gamma_X\beta_U\alpha_Z+\beta_U(\gamma+\gamma_X)\alpha_{ZX},$$

(17)

$$Bias_{Zadj}=RD_{Zadj}-\beta_X=\beta_U\alpha_X+\beta_U\gamma\alpha_{ZX}.$$

(18)

Appendix B presents the conditions for bias amplification and attenuation based on equations (17) and (18).

4.4 The BK2-Plot

The BK2-Plot provides insight into bias amplification and attenuation. To simplify the graphical display, α_{XZ} is set equal to 0. Consequently, a key determinant of whether bias amplification or bias attenuation occurs is the sign of the parameter γ_X (with α_X and α_{ZX} already specified). The

BK2-Plot in Figure 3 shows bias amplification corresponding to $\gamma_X > 0$. The BK2-Plot in Figure 4 shows bias attenuation corresponding to $\gamma_X < 0$.

The four top panels, which are based on Derivation I, are identical BK-Plots stratified by *Z*. The horizontal axis is pr(*U*=1 | *x, z*) and the vertical axis is pr(*Y*=1| *x, u*). The parallel diagonal lines represent Model Y for *X*=0 and *X*=1. The lower diagonal line plots pr(*Y*=1| *X*=0, *z*) from β to $(\beta + \beta_U)$ as a function of pr(*U*=1 | *X*=0, *z*). The upper diagonal line plots pr(*Y*=1| *X*=1, *u*) from $(\beta + \beta_X)$ to $(\beta + \beta_X + \beta_U)$ as a function of pr(*U*=1 | *X*=1, *z*). The causal effect, β_X, is the vertical distance between these parallel diagonal lines (line *X*=1 minus line *X*=0). To construct $f_{00} = \beta(1-\alpha) + (\beta + \beta_U)\,\alpha$, a vertical line is drawn at α, the intersection with *X* = 0 is determined, and a horizontal line from that intersection specifies f_{00}. To construct $f_{10} = (\beta + \beta_X)(1-\alpha-\alpha_X) + (\beta + \beta_X + \beta_U)(\alpha + \alpha_X)$, a vertical line is drawn at $(\alpha + \alpha_X)$, the intersection with *X* = 1 is determined, and a horizontal line from that intersection specifies f_{10}.

The middle panels in Figures 3 and and44 graphically link Derivations I and II by computing φ and $(\varphi + \varphi_X)$ using equations (15) and (16) with $\alpha_{XZ} = 0$. The computation of φ as a fraction γ of the distance between α and $(\alpha + \alpha_Z)$ is indicated by a dashed blue vertical line. The computation of $(\varphi + \varphi_X)$ as a fraction $(\gamma + \gamma_X)$ of the distance between $(\alpha + \alpha_X)$ and $(\alpha + \alpha_X + \alpha_Z)$ is indicated by the dashed red vertical line for $\gamma_X = 0$, and the dashed green vertical line for $\gamma_X > 0$ (Figure 3) or $\gamma_X < 0$ (Figure 4).

The bottom panels, which are based on Derivation II, are BK-Plots involving *Z*-adjusted and crude risk differences; they graphically compute $Bias_{crude}$ and $Bias_{Zadj}$ based on the results from the middle panel. To compute f_0 a blue vertical line is drawn at φ, the intersection with *X* = 0 is determined, and a horizontal line from that intersection specifies f_0. To compute f_1 a red (for $\gamma_X = 0$) or green (for $\gamma_X \neq 0$) vertical line is drawn at $(\varphi + \varphi_X)$, the intersection with *X* = 1 is determined, and a horizontal line from that intersection specifies f_1.

The causal effect β_X is the vertical difference between the diagonal lines reproduced from the top panels, namely the distance between the blue and dashed black horizontal lines, a positive quantity. RD_{crude} is the difference between f_1 and f_0 when $\gamma_X = 0$, namely the distance between red and blue horizontal lines, a negative quantity. RD_{Zadj} is the difference between f_1 and f_0 when $\gamma_X \neq 0$, namely the distance between green and blue horizontal lines, a negative quantity. Bias is indicated by the colored arrows. In Figure 3 the red downward arrow for $Bias_{Zadj}$ is larger than the green downward error for $Bias_{crude}$, indicating bias amplification. In Figure 4 the red downward arrow for $Bias_{Zadj}$ is smaller than the green downward error for $Bias_{crude}$, indicating bias attenuation.

The BK2-Plot aids intuition by explaining bias amplification and attenuation as relative shifts in vertical lines when adjusting versus not adjusting for a variable that is directly related to treatment but not directly related to outcome. In practice, it is difficult to decide *a priori* the direction and extent of the shift of vertical lines in a particular problem.

5. THE PAIRED AVAILABILITY DESIGN: THE PAD-PLOT

The paired availability design [18, 19] does not fit into a causal graph framework but can be formulated using probability theory. A brief summary is given here along with an improved version of a graphical display [18]. Let *Z* denote time period and *Y* denote outcome. Let T0 and T1 denote treatments. Treatment availability changes from time period Z=0 to time period *Z*=1. Under various assumptions that can be made more plausible by design, the causal effect of time period is

$$\Delta_{overall} = \mathrm{pr}(Y=1|Z=1) - \mathrm{pr}(Y=1|Z=0). \tag{19}$$

The goal of the paired availability design is to estimate the causal effect of treatment T1 instead of T0. Achieving this goal involves the following principal stratification model [20] and two plausible assumptions. The principal strata, denoted *R* = *r*, are based on the treatment a participant would receive if arrival was (sometimes hypothetically) in either time period are

- $R = n$, if would receive T0 regardless of time period,
- $R = c$, if would receive T0 in time period $Z = 0$ and T1 in time period $Z = 1$,
- $R = i$, if would receive T1 in time period $Z = 1$ and T0 in time period $Z = 0$,
- $R = a$, if would treatment T1 regardless of time period.

Based on the definitions, the probability of receiving T1 in a given time period is a function of the principal strata: pr(T1 | Z=0) = pr(R=i) + pr(R=n) and pr(T1 | Z=1) = pr(R=c) + pr(R=a). Therefore the effect of time period on the probability of receiving T1 is

$$\Delta_{\text{treated}} = \text{pr}(\text{T1}|Z=1) - \text{pr}(\text{T1}|Z=0) = \text{pr}(R=c) - \text{pr}(R=i). \tag{20}$$

The overall treatment effect is

$$\Delta_{\text{overall}} = \Delta_{stratum(a)}\text{pr}(R=a) + \Delta_{stratum(c)}\text{pr}(R=c) - \Delta_{stratum(i)}\text{pr}(R=i) - \Delta_{stratum(n)}\text{pr}(R=n). \tag{21}$$

where $\Delta_{stratum(r)}$ = pr(Y=1 | Z=1, r) − pr(Y=1 | Z=0, r). The following two assumptions are invoked for identifiability.

Assumption 1. The probability of outcome does not change over the time periods for R= n, a. Mathematically this assumption is pr(Y=1 | Z=0, R=n, T0) = pr(Y=1 | Z=1, R=n, T0) ≡ pr(Y=1 | n, T0) and pr(Y=1 | Z=0, R=a, T1) = pr(Y=1 | Z=1, R=a, T1) ≡ pr(Y=1 | a,T1).

Assumption 2. Under fixed availability (the increase in availability of treatment occurs at a fixed time), pr(R=i) =0. Under random availability (the increase in availability of treatment occurs at random times), receipt of T1 or T0 occurs by chance among principal strata R=c and R=i. Mathematically this assumption translates to pr(Y=1|Z=0, R=c, T0) =

pr(Y=1|Z=1, R=I, T0) ≡ pr(Y=1| c, T0) and pr(Y=1|Z=1, R=c, T1) = pr(Y=1|Z=0, R=I, T1) ≡ pr(Y=1| c, T1).

Assumption 1 implies $\Delta_{stratum(a)} = \Delta_{stratum(n)} = 0$. The addition of Assumption 2 implies $\Delta_{\text{overall}} = \Delta_{stratum(c)}$ pr(R=c) for fixed availability and $\Delta_{stratum(c)} = -\Delta_{stratum(i)}$ so $\Delta_{\text{overall}} = \Delta_{stratum(c)}$ {pr(R=c) − pr(R=i)} for random availability. This yields the well-known result

$$\Delta_{stratum(c)}=\{\mathrm{pr}(Y=1|Z=1)-\mathrm{pr}(Y=1|Z=0)\}/\{\mathrm{pr}(\mathrm{T1}|Z=1)-\mathrm{pr}(\mathrm{T1}|Z=0)\} \tag{21}$$

The PAD-Plots in Figures 5 and 6 graphically display the above equations, adding insight to the calculations.

ACKNOWLEDGEMENTS

This work was supported by the National Institutes of Health. The author thanks Jessica Myers, and the reviewers and associate editor for helpful comments.

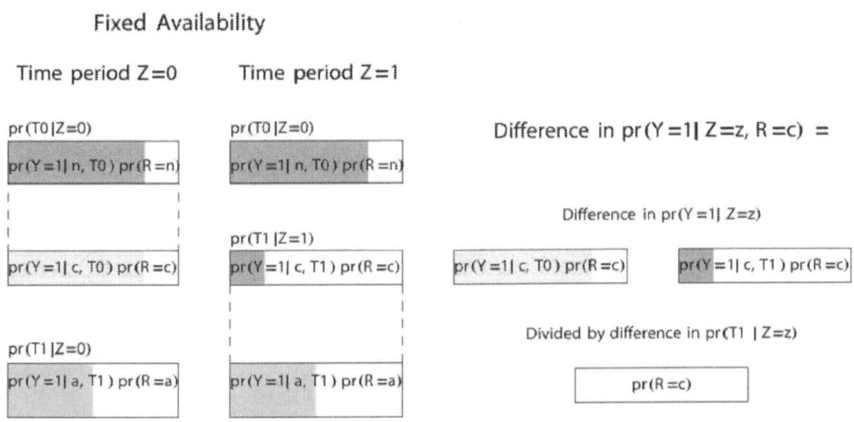

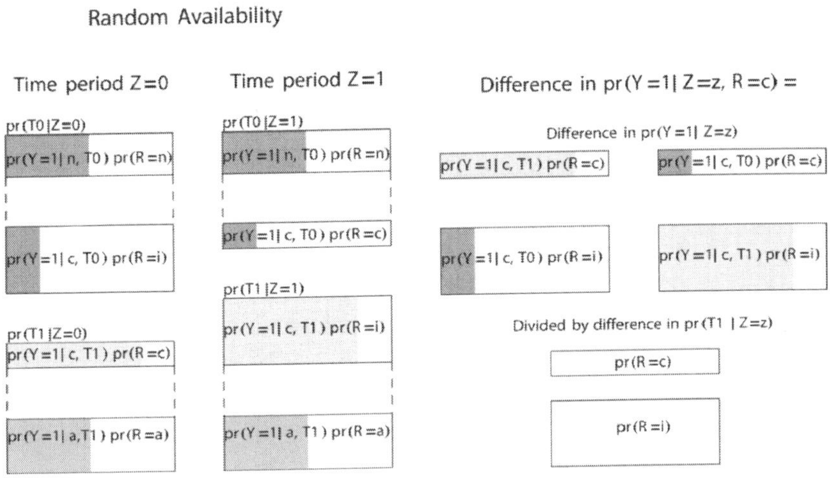

Figure 5. PAD-Plot for paired availability design. The size of each box represents pr(R=r). The size of the shaded area in the box represents pr(y|r, treatment).

REFERENCES

1. Byar DP. Why data bases should not replace randomized clinical trials. Biometrics. 1980;35:337–342.

2. Pearl J. Causality: Models, Reasoning, and Inference. 2nd ed. New York, NY: Cambridge University Press; 2009.

3. Greenland S. Quantifying biases in causal models: classical confounding vs collider-stratification bias. Epidemiology. 2003;14:300–306.

4. Shrier D. Letter to the editor. Statistics in Medicine. 2008;27:2740–2741.

5. Rubin DB. Author's reply (to Ian Shrier's Letter to the Editor) Statistics in Medicine. 2008;27:2741–2742.

6. Shrier D. Letter to the editor: Propensity scores. Statistics in Medicine. 2009;28:1317–1318.

7. Sjolander A. Letter to the editor: Propensity scores and m-structures. Statistics in Medicine. 2009;28:1416–1420.

8. Pearl J. Letter to the editor: Remarks on the method of propensity scores. Statistics in Medicine. 2009;28:1420–1423.

9. Rubin DB. Author's reply: Should observational studies be designed to allow lack of balance in covariate distributions across treatment groups? Statistics in Medicine. 2009;28:1420–1423.

10. Liu W, Brookhart MA, Schneeweiss S, Mi X, Setoguchi S. Implications of M bias in epidemiologic studies: a simulation study. Am J Epidemiol. 2012;176:938–948.

11. Bhattacharya J, Vogt WB. Do Instrumental Variables Belong in Propensity Scores? International Journal of Statistics and Economics. 2012;9:A12.

12. Pearl J. Proceedings of the Twenty-Sixth Conference on Uncertainty in Artificial Intelligence (UAI 2010) Corvallis: Association for Uncertainty in Artificial Intelligence; 2010. On a class of bias-amplifying variables that endanger effect estimates; pp. 425–432.

13. Myers JA, Rassen JA, Gagne JJ, Huybrechts KF, Schneeweiss S, Rothman KJ, Joffe MM, Glynn RJ. Effects of adjusting for instrumental variables on bias and precision of effect estimates. American Journal of Epidemiology. 2011;174:1213–1222.

14. Pearl J. Invited commentary: understanding bias amplification. American Journal of Epidemiology. 2011;174:1223–1227.

15. Myers JA, Rassen JA, Gagne JJ, Huybrechts KF, Schneeweiss S, Rothman KJ, Glynn RJ, Myers, et al. Response to "Understanding Bias Amplification". American Journal of Epidemiology. 2011;174:1228–1229.

16. VanderWeele TJ, Shpitser I. A new criterion for confounder selection. Biometrics. 2011;67:1406–1413.

17. Breslow NE, Day NE. Statistical methods in cancer research. Lyon: International Agency for Research on Cancer; 1980. p. 104.

18. Baker SG, Lindeman KS. Revisiting a discrepant result: a propensity score analysis, the paired availability design for historical controls, and a meta-analysis of randomized trials. Journal of Causal Inference. 2013 in press.

19. Baker SG, Lindeman KS. The paired availability design: a proposal for evaluating epidural analgesia during labor. Statistics in Medicine. 1994;13:2269–2278.

20. Frangakis CE, Rubin DB. Principle stratification in causal inference. Biometrics. 2002;58:21–29.

21. Gardner M. Martin Gardner's Sixth Book of Mathematical Games from Scientific American. San Francisco: W.H. Freeman and Company; 1971. p. 154.

22. Maor E. Trignometric Delights. Princeton: Princeton University Press; 1998. pp. 122–123.

23. Soto AM, Sonnenschein C, Miquel PA. On physicalism and downward causation in developmental and cancer biology. Acta Biotheor. 2008 Dec;56(4):257–274.

24. Prehn RT. Cancers beget mutations versus mutations beget cancer. Cancer Res. 1994;54:5296–5300.

25. Baker SG. Paradoxes in carcinogenesis should spur new avenues of research: An historical perspective. Disruptive Science and Technology. 2012;1:100–107.

26. Baker SG. Paradox-driven cancer research. Disruptive Science and Technology. 2013 In press.

27. Baker SG, Kramer BS. Good for women, good for men, bad for people: Simpson's paradox and the importance of sex-specific

analysis in observational studies. Journal of Women's Health & Gender-Based Medicine. 2001;10:867–872.

28. Hernán MA, Clayton D, Keiding N. The Simpson's paradox unravelled. International Journal of Epidemiology. 2011;40:780–785.

29. Arah O. The role of causal reasoning in understanding Simpson's paradox, Lord's paradox, and the suppression effect: Covariate selection in the analysis of observational studies. Emerging Themes in Epidemiology. 2008;5:5.

30. Jeon JW, Chung HY, Bae JS. Chances of Simpson's paradox. Journal of the Korean Statistical Society. 1987;16:117–125.

31. Wainer H. The BK-Plot: Making Simpson's paradox clear to the masses. Chance. 2002;15:60–62.

32. Baker SG, Freedman LS. A simple method for analyzing data from a randomized trial with a missing binary outcome. BMC Medical Research Methodology. 2003;3:8.

33. Baker SG, Kramer BS. The transitive fallacy for randomized trials: If A bests B and B bests C in separate trials, is A better than C? BMC Medical Research Methodology. 2002;2:13.

34. Baker SG, Kramer BS. Surrogate endpoint analysis: An exercise in extrapolation. Journal of the National Cancer Institute. 2013 In press.

35. Baker SG, Kramer BS. Randomized trials, generalizability, and meta-analysis: Graphical insights for binary outcomes. BMC Medical Research Methodology. 2003;3:10.

36. Gail MH, Wieand S, Piantadosie S. Biased estimates of treatment effect in randomized experiments with nonlinear regressions and omitted covariates. Biometrika. 1984;71:431–444.

37. Greenland S, Robins JM, Pearl J. Confounding and collapsibility in causal inference. Statistical Science. 1999;14:29–46.

38. Gail MH, Wacholder S, Lubin JH. Indirect corrections for confounding under multiplicative and additive risk models. American Journal of Industrial Medicine. 1988;13:119–130.

Chapter 10

The Statistical Analysis of Interval-Censored Failure Time Data with Applications

Radhey S. Singh[1#†], Dishna P. Totawattage[2]

[1]University of Waterloo, Waterloo, Canada

[2]Formerly at University of Guelph, Guelph, Canada

ABSTRACT

The analysis of survival data is a major focus of statistics. Interval censored data reflect uncertainty as to the exact times the units failed within an interval. This type of data frequently comes from tests or situations where the objects of interest are not constantly monitored. Thus events are known only to have occurred between the two observation periods. Interval censoring has become increasingly common in the areas that produce failure time data. This paper explores the statistical analysis of interval-censored failure time data with applications. Three different data sets, namely Breast Cancer,

Hemophilia, and AIDS data were used to illustrate the methods during this study. Both parametric and nonparametric methods of analysis are carried out in this study. Theory and methodology of fitted models for the interval-censored data are described. Fitting of parametric and non-parametric models to three real data sets are considered. Results derived from different methods are presented and also compared.

KEYWORDS

Interval Censoring; Survival Analysis; Parametric; Non-Parametric; Semi-Parametric; Survival Functions; Survival Curves; Kaplan-Meier Estimate; Turnbull Estimator; Logspline Estimation

1. INTRODUCTION

A great many studies in statistics deal with deaths or failures of components: they involve the numbers of deaths, the timing of death, or the risks of death to which different classes of individuals are exposed. The analysis of survival data is a major focus of statistics.

In standard time-to-event analysis, the time to a particular event of interest is observed exactly or right-censored. Numerous methods are available for estimating the survival curve and also for estimation of the effects of covariates for these cases. In certain situations, the times of the event of interest may not be exactly known. This means that it may have occurred within particular time duration. In clinical trials, patients are often seen at pre-scheduled visits but the event of interest may have occurred in between visits. These types of data are known as interval-censored data.

Right-censored data can be considered as a special case of interval-censored data. Some of the inference approaches for right-censored data can be directly, or with minor modifications, used to analyze interval-censored data. However, most of the inference approaches for right-censored data are not appropriate for interval-censored data due to fundamental differences between these two types of censoring. The censoring approach behind interval censoring is more complicated than that of right censoring. For right-censored failure time data, substantial advances in the theory and development of modern statistical methods are based on the counting processes theory, which is not applicable to interval-censored data. Due to the complexity and special structure of interval censoring, the same theory is not applicable to interval-censored data.

Interval censoring has become increasingly common in the areas that produce failure time data. Over the past two decades, a lot of literature on the statistical analysis of interval-censored failure time data has appeared.

Lindsay and Ryan [1] provided a tutorial on Biostatistical methods for interval-censored data. This paper illustrated and compared available methods which correctly treated the data as being interval-censored. This paper did not provide a full review of all existing methods. However, all approaches were illustrated on two data sets and compared with methods which ignore the interval-censored nature of the data. In this paper, we have used some of the methodologies, notations and equations used by Lindsay and Ryan [1].

Lindsay [2] showed that parametric models for interval censored data can now easily be fitted with minimal programming in certain standard statistical software packages. Regression equations were introduced and finite mixture models were also fitted. Models based on nine different distributions were compared for three examples of heavily censored data as well as a set of simulated data. It has been found that interval censoring can be ignored for parametric models. Parametric models are remarkably robust with changing distributional assumptions and more informative than the corresponding non-parametric models for heavily interval censored data.

Finkelstein and Wolfe [3] provided a method for regression analysis to accommodate interval-censored data. Finkelstein [4] develops a method for fitting proportional hazards regression model when the data contain intervalcensored observations. The method described in this paper is used to analyze data from an animal study and also a clinical trial.

Peto [5] provided a method of calculating an estimate of the cumulative distribution function from intervalcensored data, which was similar to the life-table technique.

Rosenberg [6] presented a flexible parametric procedure to model the hazard function as a linear combination of cubic B-Splines and derived maximum likelihood estimates from censored survival data. This provided smooth estimates of the hazard and survivorship functions that are intermediate between parametric and non-parametric models. HIV infections data that were intervalcensored were used to illustrate the methods.

Odell et al. [7] studied the use of a Weibull-based accelerated failure time regression model when intervalcensored data were observed. They have used two alternative methods to analyze the data. Turnbull [8] has used non-parametric estimation of a distribution function for censored data. A simple algorithm using self-consistency as a basis was used to get maximum likelihood estimates.

Farrington [9] provided a method for weak parametric modeling of interval-censored data using generalized linear models. Three types of models, namely, additive, multiplicative and proportional hazard model with discrete baseline survival function were considered. Goetghebeur and Ryan [10] introduced semi-parametric regression analysis of interval censored data. A semi-parametric approach to the proportional hazards regression analysis of interval-censored data was proposed in this paper. The method was illustrated on data from the breast cancer cosmetics trial, previously analyzed by Finkelstein [4].

Lawless [11] provides a unified treatment of models and statistical methods used in the analysis of lifetime or response time data (Chapter

3, Section 3.5.3, p. 124). Numerical illustrations and examples involving real data demonstrate the application of each method to problems in areas such as reliability, product performance evaluation, clinical trials, and experimentation in the biomedical sciences. Collet [12] describes and illustrates the modeling approach to the analysis of survival data. Some methods for analyzing interval-censored data are described and illustrated. This begins with an introduction to survival analysis and a description of four studies in which survival data was obtained. These and other data sets then illustrate the techniques presented, including the Cox and Weibull proportional hazards models, accelerated failure time models, models with time-dependent variables, interval-censored survival data and model checking.

Sun [13] has recently presented statistical models and methods specifically developed for the analysis of interval-censored failure time data. This book collects and unifies statistical models and methods that have been proposed for analyzing interval-censored failure time data. It provides the first comprehensive coverage of the topic of interval-censored data. This focuses on non-parametric and semi-parametric inferences, but it also describes parametric and imputation approaches. This paper reviews the substantial body of recent work in this field and also provides some applications.

2. STATISTICAL METHODOLOGY

2.1. Parametric Methods

The straightforward procedure to analyze censored data is to assume a parametric model for the failure times. It is possible to fit Accelerated failure time (AFT) models for a variety of distributions to interval censored data. In the statistical area of survival analysis, an accelerated failure time model is a parametric model that provides an alternative to the commonly-used proportional hazards models. A proportional hazards model assumes that the effect of a covariate is to multiply the hazard by some constant; an AFT model assumes that the effect of a covariate is to multiply the predicted event time by some constant. AFT models can therefore be framed as linear models for the logarithm of the survival time.

The results of AFT models are easily interpreted. For example, the results of a clinical trial with mortality as the endpoint could be interpreted as a certain percentage increase in future life expectancy on the new treatment compared to the control. So a patient could be informed that he would be expected to live (say) 15% longer if he took the new treatment. Hazard ratios are harder to explain in layman's terms. More probability distributions can be used in AFT models than parametric proportional hazards models. A distribution must have a parameterization that includes a scale parameter to be used in an AFT model. The logarithm of the scale parameter is then modeled as a linear function of the covariates.

The Weibull distribution (including the exponential distribution as a special case) can be parameterized as either a proportional hazards model or an AFT model, and is the only family of distributions to have this property. The results of fitting a Weibull model can therefore be interpreted in either framework. Unlike the Weibull distribution, log-logistic distribution can exhibit a non-monotonic hazard function which increases at early times and will decrease at later times.

Other distributions suitable for AFT models include the log-normal and log-gamma distributions, although they are less popular than the log-logistic, partly as their cumulative distribution functions do not have a closed form.

The SAS procedure LIFEREG provides a way of fitting accelerated failure time models for a variety of distributions to interval censored data. The AFT model is defined by the transformation

$$T_z = T_0 e^{-\beta z}, \qquad (2.1)$$

where T_z is the failure time random variable for an individual with covariate z and T_0 is the failure time that the individual would have if they had covariate value 0. The effect of changing covariates is to shrink or stretch the time to event. If β is negative, then the covariate has the effect of "speeding up time" so that individuals with larger values of z have higher failure rates and hence shorter survival times. The survival function can be written as

$$S(t;z) = P(T_z \geq t|z) = P(T_0 \geq te^{\beta z}) = S_0(te^{\beta z}), \tag{2.2}$$

where $S_0(t)$ is the survival function for an individual with covariate value 0. Taking natural logarithm, the AFT model can be expressed as

$$\log T_z = \log T_0 - \beta z. \tag{2.3}$$

If we assume that $\log T_0$ can be expressed as $\mu + \sigma W$, where W is a random variable, then the model can be written in a linear model-like form:

$$\log T_z = \mu - \beta z + \sigma W. \tag{2.4}$$

The PROC LIFEREG module of SAS fits this model, except that the sign is changed on the regression coefficients. That is, SAS fits

$$\log T_z = \mu + \beta z + \sigma W. \tag{2.5}$$

It is possible to include a variety of distributions to be placed on the error term W with SAS, including the log of the exponential, log-normal and log-gamma distributions. The intercept parameter μ and the scale parameter σ are usually not of direct interest, although for some distributions, there is a relationship between the AFT model and a proportional hazards model through the scale parameter. For example, if W is an extreme value distribution (log of a unit exponential), then T has a Weibull distribution. Note that because of the change in sign implicit in the AFT formulation, the direction of covariate effects will be opposite to those fit with a Cox proportional hazards model.

2.2. Non-Parametric Estimation of Survival Curve

2.2.1. Kaplan-Meier Estimator

The Kaplan-Meier estimator estimates the survival function for life-time data. In medical research, it might be used to measure the fraction of patients living for a certain amount of time after treatment. An economist might measure the length of time people remain unemployed after a job loss. An engineer might measure the time until failure of machine parts.

A plot of the Kaplan-Meier estimate of the survival function is a series of horizontal steps of declining magnitude which, when a large enough sample is taken, approaches the true survival function for that population. The value of the survival function between successive distinct sampled observations is assumed to be constant.

An important advantage of the Kaplan-Meier curve is that the method can take censored data into account, for instance, if a patient withdraws from a study. When no truncation or censoring occurs, the Kaplan-Meier curve is equivalent to the empirical distribution.

Let $S(t)$ be the probability that an item from a given population will have a lifetime exceeding t. For a sample from this population of size N let the observed times until death of N sample members be

$$t_1 \le t_2 \le t_3 \le \cdots \le t_N \, . \tag{2.6}$$

Corresponding to each t_i is n_i, the number "at risk" just prior to time t_i, and d_i, the number of deaths at time t_i.

Note that the intervals between each time typically will not be uniform. For example, a small data set might begin with 10 cases, have a death at Day 3, a loss (censored case) at Day 9, and another death at Day 11. Then we have $(t_1 = 3, t_2 = 11)$, $(n_1 = 10, n_2 = 8)$, and $(d_1 = 1, d_2 = 1)$.

The Kaplan-Meier estimator is the nonparametric maximum likelihood estimate of $S(t)$. It is a product of the form

$$\hat{S}(t) = \prod_{t_i < t} \frac{n_i - d_i}{n_i} \quad (2.7)$$

When there is no censoring, n_i is just the number of survivors just prior to time t_i. With censoring, n_i is the number of survivors less the number of losses (censored cases). It is only those surviving cases that are still being observed (have not yet been censored) that are "at risk" of an (observed) death.

Let T be the random variable that measures the time of failure and let $F(t)$ be its cumulative distribution function. Note that

$$S(t) = P[T > t] = 1 - P[T \le t] = 1 - F(t) \quad (2.8)$$

Consequently, the right-continuous definition of $\hat{S}(t)$ may be preferred in order to make the estimate compatible with a right-continuous estimate of $F(t)$.

With right-censored data, Kaplan and Meier [14] showed that the closed form product limit estimator is the generalized maximum likelihood estimate. This curve jumps at each observed event time.

2.2.2. Turnbull Estimator

In most applications, the data may be interval-censored. By interval-censored data, a random variable of interest is known only to lie in an interval, instead of being observed exactly. In such cases, the only information available for each individual is that their event time falls in an interval, but the exact time is unknown. A nonparametric estimate of the survival function can also be found in such interval censored situations. The survival function is perhaps the most important function in medical and health studies. In this section, the iterative procedure proposed by Turnbull [8] to estimate such function is described and illustrated.

Situations where the observed response for each individual under study is either an exact survival time or a censoring time are common in practice. Other situations, however, can occur, and amongst them we find the longitudinal studies, where the individuals are followed for a pre-fixed time period or visited periodically for a fixed number of times. In this context, the time $T_i\,(i=1,\cdots,n)$ until the occurrence of the event of interest for each individual is only known (whenever it occurs) to be within the interval between visits, i.e., between the visit in time L_i and the visit in time U_i. Note that in such studies, the survival times T_i are no longer known exactly. It is only known that the event of interest has occurred within the interval $(L_i, U_i]$ with $L_i \prec T_i \leq U_i$. Furthermore, note that if the event occurs exactly at the moment of a visit, which is very improbable but can happen, then we have an exact survival time. In this case it is assumed that $T_i = L_i = U_i$.

On the other hand, it is known for the individuals with right censoring that the event of interest did not occur until the last visit but it can happen at any time from that moment on. We therefore assumed in this case that T_i can occur within the interval (L_i, ∞) with L_i being equal to the period of time from the beginning of the study until the last visit and $U_i = \infty$.

Similarly, it is known for the individuals that are left censored, that the event of interest has occurred before the first visit and, hence, we assume that T_i falls in the interval $(0, U_i]$ with $L_i = 0$ representing the beginning of the study and U_i is the period of time from the beginning of the study until the first visit.

Note from what we have presented so far that exact survival times as well as right and left censored data, are all special cases of interval survival data with $L_i = U_i$ for exact times, $U_i = \infty$ for right censoring and $L_i = 0$ for left censoring. We can therefore state that interval survival data generalize any situation with combinations of survival times (exact or interval) and right and left censoring that can occur in survival studies.

As usual in the analysis of non-interval survival data, it is also of interest to estimate the survival function $S(t)$ and to assess the importance of potential prognostic factors. Few statistical software allow for such data, and for this reason a common practice amongst data analysts is to assume that the event occurring within the interval $(L_i, U_i]$ has occurred either at the upper/lower limit of the interval or, at the middle point of each interval. Rucker and Messerer [15], Odell et al. [7] and Dorey et al. [16] stated that assuming interval survival times as exact times can lead to biased estimates as well as results and conclusions that were not fully reliable. In this work we describe a nonparametric procedure for estimation of the survival function for interval survival data.

Peto [5] was the first to propose a non-parametric method for estimating the survival distribution based on interval-censored data. Turnbull [8] derived the same estimator, but used a different approach in estimation. Suppose $T_i (i = 1, \cdots, n)$, the survival times for n patients, are independent random variables with right continuous survival function $S(t) = \Pr(T \geq t)$. If T_i are not observed directly, but instead are known to lie in the interval $[L_i, R_i]$, then the likelihood for the n observations is,

$$L = \prod_{i}^{n} \left\{ S(L_i) - S\left(R_i^+\right) \right\} \tag{2.9}$$

By $S\left(t^+\right)$, we mean

$$\lim_{\Delta \to 0^+} S(t + \Delta) \tag{2.10}$$

which may be different from $S(t)$, since $S(t)$ is left continuous. It is important to note that different authors vary in their conventions regarding definition of the censoring interval. The Convention of Peto [5] and Turnbull [8] who assumed a closed interval, $[L_i, R_i]$, was followed. This definition facilitates the accommodation of observations that are known exactly, that is, $L_i = R_i$, but necessitates the use of the R_i^+ notation in above Equation (2.9) to allow a non-zero contribution to the likelihood for these observations. Finkelstein [4] assumed semiclosed censoring intervals, which need to add the convention that the likelihood contribution for any observation with an exact failure time, T_i, is $S(t_i)$. Good arguments against almost any convention can be made for defining the censoring intervals. In practice, the choice will have little impact and all reasonable conventions can be adopted.

Turnbull [8] derived the same estimator using an iterative self-consistency algorithm, described below. Gentleman and Geyer [17] showed that this self-consistent estimator is not always the maximum likelihood estimator (MLE), and that the MLE is not necessarily unique and discuss conditions under which this can be determined.

Since the observed event times are known to occur only within potentially overlapping intervals, the survival curve can only jump within so-called equivalence sets $\left[q_j, q_{j+1}\right]$, $j=1,\cdots,m$, where $q_j \le p_j \le q_{j+1} \le \cdots$. The curve between p_j and q_{j+1} is flat. The estimate of $S(t)$ is unique only up to these equivalence classes; any function that jumps the appropriate amount within the equivalence class will yield the same likelihood.

An analog of the Product-Limit estimator of the survival function for interval-censored data is presented in this section. This estimator, which has no closed form, is based on an iterative procedure and has been suggested by Turnbull [8].

To construct the estimator, let $0=\tau_0 \prec \tau_1 \prec \tau_2 \prec \cdots \prec \tau_m$ be a grid of time which includes all the points L_i and U_i for $i=1,\cdots,n$. For the ith observation, define a weight α_{ij} to be 1 if the interval $\left(\tau_{j-1}, \tau_j\right)$ is contained in the interval $(L_i, U_i]$ and 0, otherwise. The weight α_{ij} indicates whether the event which occurs in the interval $(L_i, U_i]$ could have occurred at τ_j. An initial guess at $S\left(\tau_j\right)$ is made and Turnbull's algorithm is as follows:

Step 1: Compute the probability of an event occurring at time τ_j by

$$p_j = S\left(\tau_{j-1}\right) - S\left(\tau_j\right), j=1,\cdots,m; \tag{2.11}$$

Step 2: Estimate the number of events which occurred at τ_j by

$$d_j = \sum_{i=1}^{n} \frac{\alpha_{ij} p_j}{\sum_{k=1}^{m} \alpha_{ik} p_k}, j=1,\cdots,m; \tag{2.12}$$

Step 3: Compute the estimated number at risk at time

τ_j by $n_j = \sum_{k=j}^{m} d_k$;

Step 4: Compute the updated Product-Limit estimator using the pseudo data found in Steps 2 and 3. If the updated estimate of S is close to the old version of S for all τ_j's, stop the iterative process, otherwise repeat Steps 1- 3, using the updated estimate of S.

2.2.3. Logspline Estimation of the Survival Curve

Kooperberg and Stone [18] have introduced the Logspline density estimation. They have developed a system for data that may be right censored, left censored, or interval censored. A fully automatic method was used to determine the estimate, which involved the maximum likelihood method and may involve stepwise knot deletion and either the Akaike information criterion (AIC) or Bayesian information criterion (BIC), was used to determine the estimate.

Kooperberg and Stone [18] provided software (logspline. fit, available through Statlib for S-plus2) which can be used to obtain smoothed estimates of the survival function based on interval censored data using splines. Smooth functions were fitted to the log-density function of the failure times within subsets of the time axis defined by the "knots", and constrained to be continuous at those points. This provides a loosely parametric framework for finding estimates of the survival and hazard functions which can be useful for exploratory data analysis. Their approach is related to that of Rosenberg [6] who uses splines to model the hazard function.

3. APPLICATIONS

There are essentially three approaches to fit survival models. The first straightforward method is the parametric approach, where a specific functional form for the baseline hazard $\lambda_0(t)$ is assumed. Examples are: models based on the exponential, Weibull, gamma and generalized F distributions. A second approach might be called a flexible or semi-parametric strategy, where mild assumptions are made about the baseline hazard $\lambda_0(t)$. Specifically, time is subdivided into reasonably small intervals and it is assumed that the baseline hazard is constant in each interval leading to a piecewise exponential model. The third approach is a non-parametric strategy that focuses on estimation of the parameters leaving the baseline hazard $\lambda_0(t)$ completely unspecified. This approach relies on a partial likelihood function proposed by Cox [19].

Five ways of estimating the time to event ignoring the effects of covariates are considered initially. Standard Kaplan-Meier estimator is used first. It is assumed that exact times of event are known and this is done either by assuming the event occurred at the left interval, or at the right interval. These two extreme cases should roughly bracket the estimates derived using the interval-censoring methods. A second approach is using a Weibull model, where the survivals function is modeled using the estimates from the SAS Proc LIFEREG. A Third procedure is to model the interval-censored nature of the data using the techniques proposed by Turnbull. The fourth is to use splines models proposed by Kooperberg and Stone [18]. Finally, the survival function is estimated using the piecewise exponential model.

3.1. Breast Cancer Data

This data set is from a retrospective study of patients with breast cancer designed to compare radiation therapy alone versus in combination with chemotherapy with respect to the time to cosmetic deterioration.

This data set has been analyzed by several authors to illustrate various methods for interval censored data. Patients were seen initially every 4 to 6 months, with decreasing frequency over time. If deterioration was seen, it was known only to have occurred between two visits. Deterioration was not observed in all patients during the course of the trial, so some data were right-censored.

The breast cancer data set is described in detail in Finkelstein and Wolfe [3] and it consists of a total of 94 observations from a retrospective study looking at the time to cosmetic deterioration. Information is available on one covariate, type of therapy, either radiation alone (coded 0), or in combination with chemotherapy (coded 1). Of the 94 observations, 56 are interval-censored and 38 are right-censored. All estimated curves for the Breast Cancer data are presented in Figure 1. KM for R and KM for L represent Kaplan Meier estimates for right censored and left censored data respectively in Figures 1-3.

3.2. AIDS Data

This data set focused on the development of drug resistance (measured using a plaque reduction assay) to zidovudine in patients enrolled in four clinical trials for the treatment of AIDS. Samples were collected on the patients at a subset of the scheduled visit times dictated by the four protocols. Since the resistance assays were very expensive, there were few assessments on each patient, resulting in very wide intervals, $[L, R]$, if resistance was seen to have occurred, and a high proportion of right-censored observations. Because of the sparseness of these data, this is a challenging data set to analyze. The variables of interest were the effects of stage of disease, dose of zidovudine and CD4 lymphocyte counts at time of randomization on the time to development of resistance. All estimated curves for the AIDS data are presented in Figure 2.

Lindsay and Ryan [1] have presented both Breast Cancer and AIDS data sets. More information about these two data sets could be found there.

Some of the analyticcal methods, notations and results presented in this paper are similar to their paper.

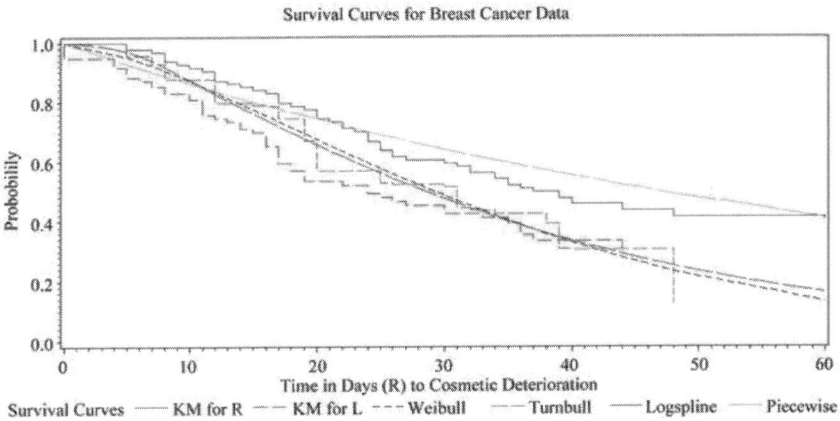

Figure 1. All estimated curves for the breast cancer data.

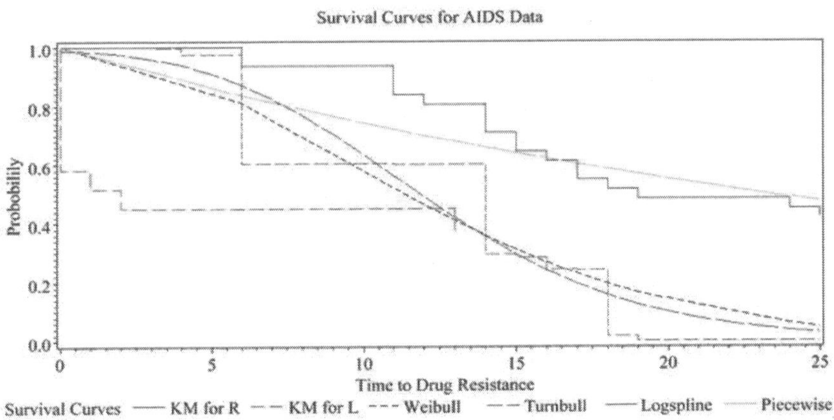

Figure 2. All estimated curves for AIDS data.

3.3. Hemophilia Data

In 1978, 262 persons with Type A or B hemophilia have been treated at Hospital Kremlin Bicetre and Hospital Coeur des Yvelines in France. Twenty-five of the hemophiliacs were found to be infected with HIV on their first test for infection. By August 1988, 197 had become infected and 43 of these had developed clinical symptoms (AIDS, lymphadenopathy, or leukopenia) relating to their HIV infection. All of the infected persons were believed to have become infected by contaminated blood factor received for their hemophilia. The observations for the 262 patients were based on a discretization of the time axis into 6-month intervals. Here time is measured in 6-month intervals, with L = 1 denoting July 1, 1978, and Z denoting chronologic time of first clinical symptom. The 25 hemophiliacs infected at entry are assigned L = 1. Victor and Stephen [20] have presented this data in their paper and more information could be found there. They have initially analyzed Hemophilia data considering this as a Doubly-Censored Survival Data. However, we have analyzed this data set taking left censoring, right censoring and interval censoring into consideration. All estimated curves for the Hemophilia data are presented in Figure 3.

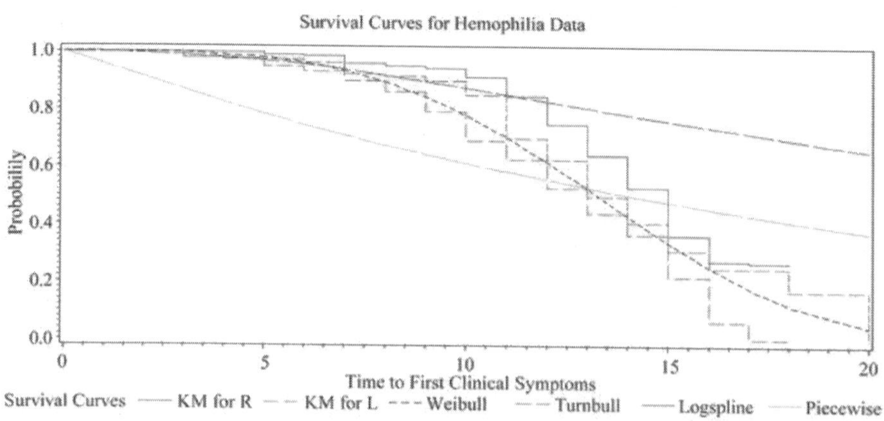

Figure 3. All estimated curves for hemophilia data.

For the breast cancer example, the Kaplan-Meier estimates, bracket the Turnbull estimate. The Turnbull curve lies very close to both Weibull and logspline curves. At the same time, estimates from the Weibull, and logspline estimates are quite close to each other. The piecewise curve does not properly fall within the Kaplan-Meier estimates.

The estimated survival curve for the AIDS data took very few steps in the non-parametric models, which reflected the high degree of censoring in this small data set. The Kaplan-Meier estimates no longer bracketed the Turnbull estimate, mainly because the Turnbull estimate had very few jumps due to the particular configuration of this data set. The logspline estimate also tracked the parametric models closely. The non-parametric methods were not very helpful in understanding the AIDS data.

For the Hemophilia data, the results were quite similar to breast cancer data except the logspline model. Results derived for the piecewise exponential are not accurate for all three data sets.

3.4. Covariate Effects on Time to Event

To compare the two treatments, for Breast cancer data a retrospective study of 46 radiation only and 48 radiation plus chemotherapy patients was conducted. Using Turnbull's algorithm the estimated survival functions were obtained for radiotherapy only and radiation plus chemotherapy groups respectively, which are shown in Figure 4. Note that the estimated survival curves did not show striking differences from 0 to 18 months. From 18 onwards, however, a fast decay of the curve is seen for patients given radiotherapy plus chemotherapy. Note, for instance, that only 11.06% of the patients in the radiotherapy plus chemotherapy group were estimated to be free of any evidence of breast retraction at time t = 40 months against 47.37% in the radiotherapy group.

Using the midpoint of each interval, is a common practice among analysts due to the lack of well-known statistical methodology and available software. Then applying the Kaplan-Meier method, we obtained the estimated survival curves presented in Figure 5. The curves estimated previously are also shown in the graph. Comparing the curves we can see that the estimates obtained using both, the midpoints and the intervals, are very similar to each other at several times but they tend to be under or over estimated at others. Although not shown here, under or over estimation became more evident if it is assumed that the event occurred to the end or at the beginning of each interval instead of at the midpoint. The range of each interval also contributes for the magnitude of these differences. They are more accentuated as the range of each interval increases.

Positive parameter estimates in Cox regression indicate higher failure rates for individuals with larger values of the covariate. The exponential model parameter should be of comparable magnitude to the Cox model, but with the sign reversed. The Weibull is the only family of models that is both proportional hazards and AFT. It can easily be shown that the estimated regression coefficient should be comparable to the coefficients from the Cox.

Results using the Cox regression models assuming exact event times (taken to be the left, midpoint and right extremes of the interval), and based on the exponential, Weibull and log-normal models for the breast cancer data is shown in Table 1. Parameter estimates of each of the above models considered, standard errors and P-values obtained for these different models are presented in this Table. All four analyses give similar results about the treatment comparison and suggest that the adjuvant chemotherapy significantly increases the risk of breast retraction. Note that the Cox analysis has the minimal impact from the differing assumptions about timing of events.

Cox models are fitted with left end point, middle point and right end point. Results obtained for fitting these models for the AIDS data with all four covariates, stage, dose, CD41 and CD42 are shown in Table 2.

The results differed from those obtained from the Breast Cancer data, although none of the estimated covariate effects are significant except in two instances. Possible explanations are that the sample size is quite small, and the observed information is very limited due to interval-censoring. Another reason could be that the covariates are correlated. The last two covariates, the indicators of CD4 count, CD41 and CD42 are correlated and both are also correlated with the other covariates the stage of the disease. Therefore, the last two covariates CD41 and CD42 are removed, and the new analysis results are presented in Table 3 along with the other results for exponential, Weibull, Log-Normal and piecewise exponential.

Now we take a look at the individual effects of stage and dose on the time to development of resistance. For stage, all methods indicate an increased risk of developing resistance for the patients in a later stage of disease. The non-parametric methods do not perform as well as the parametric methods. Unlike the breast cancer data, changing assumptions about when events are assumed to occur has a big impact on the Cox analysis. The strength of the significance is also affected when the midpoint or right extreme of the interval is used as the exact event time. With such large effects, using a method which accounts for the interval-censored nature of the data is preferable, but with so few steps in the survival curve using the non-parametric methods, a parametric analysis is the best choice. Similar trends are seen in fitting the effect of dose. The Cox model results are highly dependent on the assumptions about when the event oc-Interval-censored data often occur in medical applications. As seen in the AIDS data set, when data are heavily censored, making assumptions about when events occurred and using techniques such as Cox regression can lead to inaccurate conclusions. It can also result in unstable estimation in the non-parametric methods. The parametric methods available in SAS, S-Plus and R are the most readily available alternatives. As seen with the examples presented in this paper, these parametric approaches can be highly satisfactory in their performance. This is especially so if one chooses the Weibull or lognormal family that allows a reasonably wide range of distributional shapes.

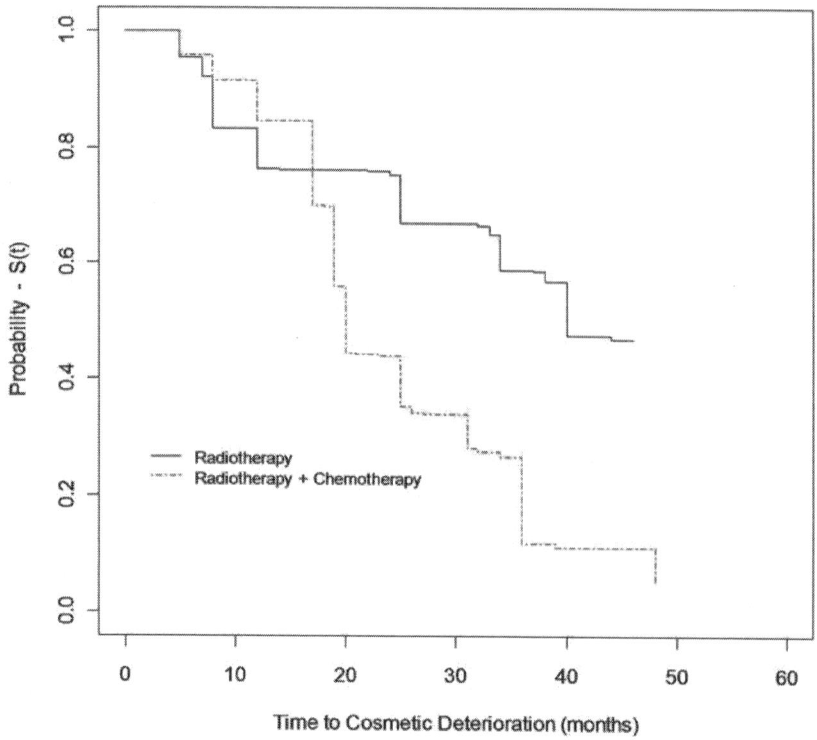

Figure 4. Estimated survival based on interval-censored data: Breast cancer data.

Results using the Cox regression models, the exponential, Weibull and log-normal models for the Hemophilia data are shown in Table 4. Parameter estimates of each of the above models considered, standard errors and Pvalues obtained for these different models are presented in this Table. All of these four analyses show significant evidence of treatment effects.

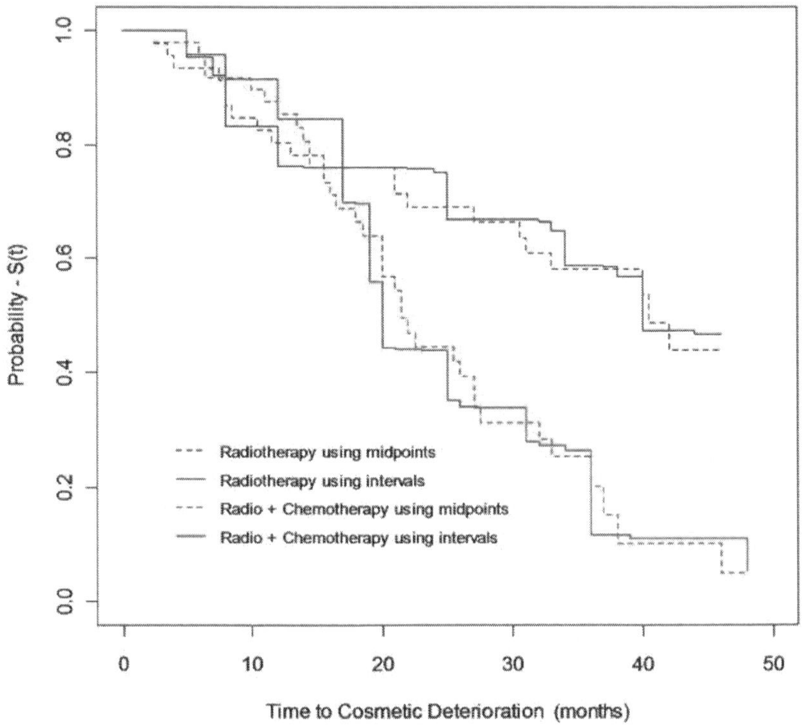

Figure 5. Estimated survival functions using midpoints and intervals: Breast cancer data.

Table 1. Breast cancer data: Effect of therapy on time to event.

Type of Model	Parameter Estimate of Model	Standard Error of Parameter Estimate	P-Value
Cox (left)	0.912	0.287	0.001
Cox (midpoint)	0.900	0.285	0.001
Cox (right)	0.769	0.285	0.007
Exponential	−0.742	0.277	0.006
Weibull (theraphy)	−0.568	0.176	<0.001
Weibull (scale)	0.619	0.074	
Lognormal (theraphy)	−0.421	0.203	0.037
Lognormal (scale)	0.882	0.097	

Table 2. Aids data: Effect of stage of disease, dose of zidovudine, CD41 and CD42.

Type of Model	Parameter Estimate of Model	Standard Error of Parameter Estimate	P-Value
Stage of Disease			
Cox (left)	0.8754	0.6558	0.1819
Cox (midpoint)	1.0083	0.6236	0.1059
Cox (right)	0.0198	0.7281	0.9784
Dose of Zidovudine			
Cox (left)	0.6394	0.7987	0.4234
Cox (midpoint)	0.1230	0.7365	0.8673
Cox (right)	0.0215	0.6440	0.9733
100 < CD4 < 399 ($CD4_1$)			
Cox (left)	0.1447	0.8027	0.8570
Cox (midpoint)	−1.3723	0.9036	0.1288
Cox (right)	−2.1488	1.0259	0.0362
CD4 > 400 ($CD4_2$)			
Cox (left)	0.1459	0.8675	0.8664
Cox (midpoint)	−1.6083	0.9181	0.0798
Cox (right)	−2.3043	0.9016	0.0106

Table 3. Aids data: Effect of stage of disease and dose of zidovudine.

Model	Estimate	Standard Error	P-Value
Stage of Disease			
Cox (left)	0.7923	0.4928	0.1079
Cox (midpoint)	1.3980	0.4991	0.0051
Cox (right)	0.7708	0.5148	0.1343
Exponential	−1.3076	0.5139	0.0109
Weibull	−0.7185	0.2711	0.0080
(Weibull scale)	0.3934	0.1389	
Log-Normal	−0.8467	0.2413	0.0004
(Log-Normal scale)	0.3880	0.1302	
Piecewise			
Dose of Zidovudine			
Cox (left)	0.5672	0.5368	0.2907
Cox (midpoint)	0.5333	0.5839	0.3611
Cox (right)	0.3132	0.5544	0.5722
Exponential	−0.5843	0.5431	0.2820
Weibull	−0.3527	0.2845	0.2151
Weibull (scale)	0.4898	0.1889	
Log-Normal	−0.3920	0.3748	0.2956
Log-Normal (scale)	0.7185	0.2563	

4. CONCLUSIONS AND FURTHER WORK

Interval censoring has become increasingly common in the areas that produce failure time data. This type of data frequently comes from tests or situations where the objects of interests are not constantly monitored. Thus events are known only to have occurred within particular time durations. The purpose of this study was to illustrate available parametric and non-parametric methods that consider the data as being interval censored.

The time to event ignoring the effects of covariates has been considered using five different techniques. The Kaplan-Meier estimator is used, assuming the event occurred at the left interval, or at the right interval. A Weibull model, where the survival function is modeled using the estimates from SAS PROC LIFEREG is carried out. A Third approach is accomplished by modeling the interval-censored nature of the data using the methods proposed by Turnbull [8]. Splines models presented by Kooperberg and Stone [18] are implemented. Finally, the survival function is estimated using the piecewise exponential model. Parametric models for interval-censored data can follow a number of distributions such as generalized gamma, the log-normal, the Weibull and the exponential distribution. Different independent covariates or categorical variables have also been included in the model to study their effect on the response variable.

Parametric and non-parametric methods of analysis are two different types of techniques in general for the analysis of censored data. However, for the analysis of interval-censored data, the terminology behind them is the same. An important advantage of parametric inference approaches is that their implementation is quite straightforward in principle and the theory of standard maximum likelihood can be applied. A primary disadvantage of these methods is that there often does not exist enough prior information or data to verify a parametric model. The major advantage of non-parametric methods such as Kaplan-Meier and Turnbull approach is that, one can avoid complicated interval censoring issues and make use of the existing inference procedures for the rightcensored data. It is also assumed that the censoring mechanism or variables are independent of the survival variables of interest.

For the Breast Cancer example, the Kaplan-Meier estimates, bracket the Turnbull estimate. The Turnbull curve lies very close to both Weibull and logspline curves. At the same time, Weibull estimates and logspline estimates are quite close to each other. Cox regression models and parametric models with covariates using Exponential, Weibull and Lognormal were fitted. Four analyses produced similar results qualitatively and all showed an increased hazard for group on radiation and chemotherapy which was statistically significant.

Unlike the results obtained for the Breast Cancer data, the estimated survival curve for AIDS data took very few steps in the non-parametric models. Possible explanations could be that the sample size is small, and the observed information was very limited due to interval censoring. The Kaplan-Meier estimates no longer bracketed the Turnbull estimate. The logspline estimate tracks the parametric models closely. The non-parametric methods were not very helpful in understanding the AIDS data. For the covariate effects of time to development of resistance, stage and dose were taken into account. All methods indicated an increased risk of developing resistance for the patients in a later stage of disease. The Cox model results were highly dependent on the assumptions about when the event occurred. No methods showed a significant effect of dose on the time to development of resistance.

The estimated survival curve for the AIDS data took very few steps in the non-parametric models, which reflected the high degree of censoring in this small data set. For the AIDS data, we took a look at the individual effects of stage and dose on the time to development of resistance. For these data, the non-parametric methods were not very helpful in understanding this data. For the breast cancer data, the four analyses gave similar results qualitatively. All showed an increased hazard for the group on radiation and chemotherapy which was statistically significant. Kaplan-Meier estimates, as expected, bracket all other survival curves.

Table 4. Hemophilia data: Effect of treatment on time to event.

Type of Model	Parameter Estimate of Model	Standard Error of Parameter Estimate	P-Value
Cox (left)	−0.66678	0.14612	<0.0001
Cox (midpoint)	−0.73541	0.14594	<0.0001
Cox (right)	−0.88687	0.14645	<0.0001
Exponential	0.5515	0.1455	0.0002
Weibull (trt)	0.2240	0.0389	<0.0001
Weibull (scale)	0.2566	0.0188	
Lognormal (trt)	0.2313	0.0496	<0.0001
Lognormal (scale)	0.3583	0.0233	

For the Hemophilia data, the results derived were very much similar to those of Breast Cancer and AIDS data sets except the logspline model. Cox regression models and parametric models with covariates using Exponential, Weibull and Lognormal were fitted. Four analyses produced similar results qualitatively and all have shown an increased hazard for the group on radiation and chemotherapy, which is statistically significant. Results derived for the piecewise exponential are not accurate for all three data sets.

Major statistical packages such as SAS, S-Plus and R have procedures for analyzing interval-censored data using parametric models. Some non-parametric methods are easily programmed. In particular, the Turnbull [8] method for non-parametric estimation of the survival distribution, the Kooperburg and Stone [18] logspline estimates of the survival function and Finkelstein's (1986) test for covariates are recommended. As seen in the AIDS data set, when data are heavily censored, making assumptions about when events occurred and using techniques such as Cox regression can lead to inaccurate conclusions. It can also result in unstable estimation in the non-parametric methods.

As examples presented in this paper show, parametric approaches can be highly satisfactory in their performance. Especially when the Weibull or log-normal family is chosen, it allows a reasonably wide range of distributional shapes. To allow more flexible modeling with weak parametric assumptions, we suggest the use of a piecewise constant hazards model. Finkelstein and Wolfe [3], Self and Grossman [21], Miller [22] and Buckley and James [23] have all proposed tests for assessing the covariate effects.

REFERENCES

1. J. C. Lindsey and L. M. Ryan, "Tutorial in Biostatistics Methods for Interval-Censored Data," Statistics in Medicine, Vol. 17, No. 2, 1998, pp. 219-238.

2. J. K. Lindsey, "A Study of Interval Censoring in Parametric Regression Models," Life Time Data Analysis, Vol. 4, No. 4, 1998, pp. 329-354.

3. D. M. Finkelstein and R. A. Wolfe, "A Semi-Parametric Model for Regression Analysis of Interval Censored Failure Time Data," Biometrics, Vol. 41, No. 4, 1985, pp. 933- 945.

4. D. M. Finkelstein, "A Proportional Hazards Model for Interval-Censored Failure Time Data", Biometrics, Vol. 42, No. 4, 1986, pp. 845-854.

5. R. Peto, "Experimental Survival Curves for Interval-Censored Data," Applied Statistics, Vol. 22, No. 1, 1973, pp. 86-91

6. P. S. Rosenberg, "Hazard Function Estimation Using BSplines," Biometrics, Vol. 51, No. 3, 1995, pp. 874-887. doi:10.2307/2532989

7. P. M. Odell, K. M. Anderson and R. B. Agostino, "Maximum Likelihood Estimation for Interval-Censored Data Using a Weibull-Based Accelerated Failure Time Model," Biometrics, Vol. 48, No. 3, 1992, pp. 951-959.

8. B. W. Turnbull, "The Empirical Distribution Function with Arbitrarily Grouped, Censored and Truncated Data", Journal of the Royal Statistical Society, Series B, Vol. 38, No. 3, 1976, pp. 290-295.

9. C. P. Farrington, "Interval Censored Survival Data: A Generalized Linear Modeling Approach," Statistics in Medicine, Vol. 15, No. 3, 1996, pp. 283-292.

10. E. Goetghebeur and L. Ryan, "Semi-Parametric Regression Analysis of Interval-Censored Data," Biometrics, Vol. 56, No. 4, 2000, pp. 1139-44.

11. J. F. Lawless, "Statistical Models and Methods for Lifetime Data," Wiley, 2003.

12. D. Collett, "Modelling Survival Data in Medical Research," Chapman and Hall, London, 2003.

13. J. Sun "The Statistical Analysis of Interval-Censored Failure Time Data," Springer, New York/Heidelberg, 2006.

14. E. L. Kaplan and P. Meier, "Nonparametric Estimation from Incomplete Observations," Journal of the American Statistical Association, Vol. 53, No. 282, 1958, pp. 457- 481.

15. G. Rucker and D. Messerer, "Remission Duration: An Example of Interval-Censored Observations," Statistics in Medicine, Vol. 7, No. 11, 1988, pp. 1139-1145.

16. F. J. Dorey, R. J. Little and N. Schenker, "Multiple Imputation for Threshold-Crossing Data with Interval Censoring", Statistics in Medicine, Vol. 12, No. 17, 1993, pp. 1589-1603. doi:10.1002/sim.4780121706

17. R. Gentleman and C. J. Geyer, "Maximum Likelihood for Interval Censored Data: Consistency and Computation," Biometrika, Vol. 81, No. 3, 1994, pp. 618-623.

18. C. Kooperberg and C. J. Stone, "Logspline Density Estimation for Censored Data," Journal of Computational and Graphical Statistics, Vol. 1, No. 4, 1992, pp. 301- 328.

19. D. R. Cox, "Regression Models and Life Tables (with Discussion)," Journal of the Royal Statistical Society, Series B, Vol. 34, No. 2, 1972, pp. 187-220.

20. V. De Gruttola and S. W. Lagakos, "Analysis of DoublyCensored Survival Data, with Application to AIDS," Biometrics, Vol. 45, No. 1, 1989, pp. 1-11.

21. S. G. Self and E. A. Grossman, "Linear Rank Tests for Interval-Censored Data with Application to PCB levels in Adipose Tissue of

Transformer Repair Workers," Biometrics, Vol. 42, No. 3, 1996, pp. 521-530.

22. R. G. Miller, "Least Squares Regression with Censored Data," Biometrika, Vol. 63, No. 3, 1976, pp. 447-464.

23. J. Buckley and I. James, "Linear Regression with Censored Data," Biometrika, Vol. 66, No. 3, 1979, pp. 429-436.

Chapter 11

Why Well Spread Probability Samples Are Balanced

Anton Grafström[1], Niklas L. P. Lundström[2]

[1]Department of Forest Resource Management, Swedish University of Agricultural Sciences, Umeå, Sweden

[2]Department of Mathematics and Mathematical Statistics, Umeå University, Umeå, Sweden

ABSTRACT

When sampling from a finite population there is often auxiliary information available on unit level. Such information can be used to improve the estimation of the target parameter. We show that probability samples that are well spread in the auxiliary space are balanced, or approximately balanced, on the auxiliary variables. A consequence of this balancing effect is that the Horvitz-Thompson estimator will be a very good estimator for any target variable that can be well approximated by a Lipschitz continuous function of the auxiliary variables. Hence we give a theoretical motivation for use of well spread probability samples. Our conclusions imply that well spread samples, combined with the HorvitzThompson estimator, is a good strategy in a varsity of situations.

KEYWORDS

Balanced Sample; Local Pivotal Method; Spatial Balance; Spatially Correlated Poisson Sampling; Voronoi Polytopes

1. INTRODUCTION

In many fields there has been a great interest in selecting samples that are well spread or spatially balanced. Such samples are considered to produce good estimates for target variables that exhibit spatial trends, see e.g. [1,2]. The focus in this paper is to explain the connection between a well spread sample and a balanced sample. Roughly speaking, a sample is well spread if the number of selected units is close to what is expected on average, in every part of the auxiliary space. A sample is balanced on a variable if the Horvitz-Thompson (HT) estimator of the total of that variable agree exactly with the known population total of the variable. In fact, with a short analysis, this paper clarifies why a well spread sample is approximately balanced. We also explain that, if the sample is well spread, the variance of commonly used estimators is usually low.

It is well known that samples that are balanced or approximately balanced on the auxiliary variables may be selected by using the cube method, see e.g. [3]. Sampling methods for selection of well spread samples in a general auxiliary space, by utilizing a distance function, are more recent and less well known than the cube method. Two such methods are the local pivotal method (LPM) and spatially correlated Poisson sampling (SCPS). The LPM design, based on the pivotal method [4], was first introduced in [5]. The other method, SCPS, was first introduced in [6] and it is a special case of the method described in [7].

In many areas, such as forest inventories, environmental studies, and even in official statistics, different forms of stratification are commonly used to obtain samples that are well spread geographically or in other available information. Often, stratification is used as a variance reduction technique without particular interest in the different strata. Constructing a stratified sampling design is often not straightforward, especially if several mixed auxiliary variables are available. It is not uncommon that statisticians try to stratify using several variables, but crossing all strata of all variables usually results in cells that are too small. In such situations it may be preferable and less complicated to define a distance measure in the auxiliary space, and then use a sampling method that in general avoid selection of nearby units, thus forcing the sample to be well spread.

In Section 2, a theoretical motivation for the balancing effect of well spread samples is given. In Section 3, we give arguments indicating that using well spread samples provides a small anticipated variance for the HT-estimator under a very general super-population model. Some sampling methods for selecting well spread samples are briefly discussed in Section 4. Final comments are **provided in Section 5.**

2. MAIN RESULTS

We start by introducing some notation and assumptions. Let $U=\{1,2,\cdots,N\}$ be a population of N units. We wish to select a probability sample s of size n in order to estimate some characteristics of U. It is assumed that we have access to auxiliary information on unit level, i.e. the values of q auxiliary variables $\boldsymbol{x}_i=\left(x_{i1},x_{i2},\cdots,x_{iq}\right)\in R^q$ are known for each unit $i\in U$. We also assume that it is possible to calculate the distance $d\left(\boldsymbol{x}_i,\boldsymbol{x}_j\right)$ between two units i and j in the auxiliary space. Usually the total

$Y = \sum_{i \in U} y_i$ of one or more target variables are the parameters we wish to estimate. It is assumed that each population unit i is included in the sample with a known probability π_i, $i \in U$, with $\sum_{i \in U} \pi_i = n$, where n is the sample size. In this case the unbiased and commonly used HT-estimator [8] of Y is

$$\hat{Y} = \sum_{i \in s} \frac{y_i}{\pi_i}.$$

We are now ready to formalize what a well spread sample is. As suggested in [2], we use Voronoi polytopes to measure how well spread a sample is. The Voronoi polytope p_i, for $i \in s$, includes all population units j satisfying $d(x_i, x_j) \leq d(x_k, x_j)$ for all sample units $k \in s$. Let ni denote the number of units in pi, with the correction that if a unit j is included in mj polytopes, then j is counted as $1/m_j$. Next, let vi be the sum of the inclusion probabilities in pi. Again, if a unit j is included in mj polytopes, then its inclusion probability is divided equally (π_j/m_j) to each of the mj polytopes. Hence,

$$n_i = \sum_{j \in p_i} \frac{1}{m_j} \text{ and } v_i = \sum_{j \in p_i} \frac{\pi_j}{m_j}.$$

Note that $\sum_{i \in s} n_i = N$ and $\sum_{i \in s} v_i = n$. We are now ready to give the definition of a well spread sample.

Definition 1 A sample is said to be well spread (or spatially balanced) with respect to the inclusion probabilities if each vi is equal or close to 1.

As a measure of how well spread a sample is, we may use

$$B = \frac{1}{n}\sum_{i \in s}(v_i - 1)^2, \qquad (1)$$

see e.g. [2]. A small value of B indicates a very well spread sample. The mean of B over repeated samples is an indicator of how well spread samples a design produces. We next define a balanced and an approximately balanced sample.

Definition 2 We say that a sample s is balanced on the auxiliary x-variables if

$$\sum_{i \in U} \mathbf{x}_i = \sum_{i \in s} \frac{\mathbf{x}_i}{\pi_i}.$$

Moreover, a sample is said to be approximately balanced if $\sum_{i \in U} \mathbf{x}_i$ is close to $\sum_{i \in s} \frac{\mathbf{x}_i}{\pi_i}$.

In order to show that well spread samples are balanced we start by making three quite strong assumptions on the sample and the population. Later we will relax these assumptions a bit and then show that a larger class of well spread samples are approximately balanced. We first assume the following.

(A.0) In each polytope the inclusion probabilities sum to 1, i.e.

$$v_i = 1 \text{ for all } i \in s.$$

(A.1) In each polytope the inclusion probabilities are equal, i.e. for every $i \in s$, we assume

$$\pi_j = \pi_i \text{ for all } j \in p_i.$$

(A.2) In each polytope the auxiliary variables are equal for all units, i.e. for every $i \in s$,

$$d(\boldsymbol{x}_i, \boldsymbol{x}_j) = 0 \text{ for all } j \in p_i.$$

The assumptions (A.0) and (A.1) tell us that the size ni of the polytope p_i is equal to $1/\pi_i$ and (A.2) tells us that $f(\boldsymbol{x}_j) = f(\boldsymbol{x}_i)$, for $j \in p_i$. Note that π_j, $\boldsymbol{x}_j$ and hence $f(\boldsymbol{x}_j)$ are allowed to vary between polytopes. Under the three assumptions it follows that

$$\sum_{i \in U} f(\boldsymbol{x}_i) = \sum_{i \in s} \sum_{j \in p_i} \frac{f(\boldsymbol{x}_j)}{m_j}$$
$$= \sum_{i \in s} f(\boldsymbol{x}_i) \sum_{j \in p_i} \frac{1}{m_j} = \sum_{i \in s} \frac{f(\boldsymbol{x}_i)}{\pi_i}. \tag{2}$$

Thus the sample is balanced on any function $f(x)$ and in particular, it is balanced on the auxiliary variables if we put $f(\boldsymbol{x}) = \boldsymbol{x}$.

The next step consists of introducing the following three new and less restrictive assumptions.

$(A'.1)$ For each polytope pi, the inclusion probabilities satisfies

$$|\pi_i - \pi_j| \le \gamma_i \text{ for all } j \in p_i \text{ and some } 0 \le \gamma_i < \pi_i.$$

$(A'.2)$ In each polytope p_i, we have

$$d(\boldsymbol{x}_i, \boldsymbol{x}_j) \le r_i \text{ for all } j \in p_i \text{ and some } r_i \ge 0.$$

$(A'.3)$The target is a Lipschitz continuous function of the auxiliary variables, i.e.

$$\left|f(\boldsymbol{x}_i)-f(\boldsymbol{x}_j)\right|\le c_i d(\boldsymbol{x}_i,\boldsymbol{x}_j)$$

for all $j \in p_i$ and some $c_i \ge 0$.

Remark 1 Concerning the validity of assumption $(A'.1)$, the inclusion probabilities are (if unequal) supposed to be derived from the auxiliary x-variables, perhaps they are chosen proportional to one of the x-variables, so they should not vary much within a polytope. Remember that the polytopes are constructed by grouping together units with similar x-values.

We are now ready to state and prove our main result.

Theorem 1 Let S be a well spread sample satisfying $|v_i - 1| \le \epsilon_i$ for all $i \in S$ and for some $\epsilon_i \ge 0$. Assume also that S is from a population satisfying assumptions $(A'.1)$-$(A'.3)$. Then S is approximately balanced. In particular,

$$\sum_{i\in s}\frac{f(\boldsymbol{x}_i)-c_i r_i}{\pi_i+\gamma_i}(1-\epsilon_i)\le\sum_{i\in U}f(\boldsymbol{x}_i)$$
$$\le\sum_{i\in s}\frac{f(\boldsymbol{x}_i)+c_i r_i}{\pi_i-\gamma_i}(1+\epsilon_i).$$

By sending $\epsilon_i, c_i r_i, \gamma_i \to 0$ we obtain exact balance on the target since

$$\sum_{i\in s}\frac{f(\boldsymbol{x}_i)}{\pi_i}=\sum_{i\in U}f(\boldsymbol{x}_i).$$

Note also that if we put $f(\boldsymbol{x})=\boldsymbol{x}$ then we get a balanced sample, see Definition 2. Besides that Theorem 1 tells us that when $\epsilon_i, r_i c_i, \gamma_i$ are small, the sample will be approximately balanced on $f(x)$ and x, it also gives bounds for the target parameter $\sum_{i \in U} f(x_i)$. We can however do better than the bounds in Theorem 1. For instance, we have

$$\sum_{i \in s} \sum_{j \in p_i} \frac{f(x_i) - c_i d(x_i, x_j)}{m_j} \leq \sum_{i \in U} f(x_i)$$

$$\leq \sum_{i \in s} \sum_{j \in p_i} \frac{f(x_i) + c_i d(x_i, x_j)}{m_j},$$

but these bounds are constructed by applying a worst case senario, within each polytope, so we cannot expect the bounds to be very good.

Proof of Theorem 1. By assumption $(A'.1)$ and since $|v_i - 1| \leq \epsilon_i$ we have, for all $i \in s$,

$$1 + \epsilon_i \geq \sum_{j \in p_i} \frac{\pi_j}{m_j} \geq n_i (\pi_i - \gamma_i) \text{ and}$$

$$1 - \epsilon_i \leq \sum_{j \in p_i} \frac{\pi_j}{m_j} \leq n_i (\pi_i + \gamma_i). \tag{3}$$

The inequalities in (3) give

$$\frac{1 - \epsilon_i}{\pi_i + \gamma_i} \leq n_i \leq \frac{1 + \epsilon_i}{\pi_i - \gamma_i}. \tag{4}$$

Moreover, from assumptions $(A'.2)$ and $(A'.3)$ we see that

$$|f(x_j) - f(x_i)| \leq c_i d(x_i, x_j) \leq c_i r_i. \tag{5}$$

Theorem 1 now follows from (4) and (5). In particular,

$$\sum_{i\in U} f(\boldsymbol{x}_i) = \sum_{i\in s}\sum_{j\in p_i} \frac{f(\boldsymbol{x}_j)}{m_j} \geq \sum_{i\in s} n_i \left(f(\boldsymbol{x}_i) - c_i r_i\right)$$
$$\geq \sum_{i\in s} \frac{f(\boldsymbol{x}_i) - c_i r_i}{\pi_i + \gamma_i}(1 - \epsilon_i)$$

and

$$\sum_{i\in U} f(\boldsymbol{x}_i) = \sum_{i\in s}\sum_{j\in p_i} \frac{f(\boldsymbol{x}_j)}{m_j} \leq \sum_{i\in s} n_i \left(f(\boldsymbol{x}_i) + c_i r_i\right)$$
$$\leq \sum_{i\in s} \frac{f(\boldsymbol{x}_i) + c_i r_i}{\pi_i - \gamma_i}(1 + \epsilon_i).$$

The proof is complete.

3. VARIANCE UNDER A GENERAL MODEL

It is interesting to see how well spread samples perform under a general super-population model. Following [9], but here with a possibly non-linear model, we assume

$$y_i = f(\boldsymbol{x}_i) + \epsilon_i \text{ for all } i \in U, \tag{6}$$

where $f(\cdot)$ is a Lipschitz continuous function, $E_M(\epsilon_i) = 0$, $V_M(\epsilon_i) = \sigma^2(\boldsymbol{x}_i) = \sigma_i^2$, and $\sigma^2(\cdot)$ is a Lipschitz continuous function. Moreover,

$$\operatorname{cov}_M(\epsilon_i, \epsilon_j) = \sigma_i \sigma_j \rho_{ij} \text{ with } i \neq j,$$

where $E_M(\cdot)$, $V_M(\cdot)$ and $\operatorname{cov}_M(\cdot,\cdot)$ are the expectation, variance and covariance under the model. The correlations ρ_{ij} are supposed to be decreasing in function of the distance between the units i and j.

With some routine calculations, the anticipated variance of the HT-estimator under model (6) can be shown to be

$$E_pE_M\left(\hat{Y}-Y\right)^2 = E_p\left(\sum_{i\in s}\frac{f(x_i)}{\pi_i}-\sum_{i\in U}f(x_i)\right)^2 + \sum_{i\in U}\sum_{j\in U}\sigma_i\sigma_j\rho_{ij}\frac{\pi_{ij}-\pi_i\pi_j}{\pi_i\pi_j}, \quad (7)$$

where $E_p(\cdot)$ is the expectation under the design. Now, if we study expression (7), it becomes evident that we want the samples to be as balanced as possible on $f(x)$ to minimize the term

$$E_p\left(\sum_{i\in s}\frac{f(x_i)}{\pi_i}-\sum_{i\in U}f(x_i)\right)^2.$$

We also want to make sure that π_{ij} is small whenever ρ_{ij} is large in order to minimize the term

$$\sum_{i\in U}\sum_{j\in U}\sigma_i\sigma_j\rho_{ij}\frac{\pi_{ij}-\pi_i\pi_j}{\pi_i\pi_j}. \quad (8)$$

If the samples are selected to be well spread (i.e. small joint inclusion probabilities for nearby units), then both terms in (7) becomes small. However, if the model standard deviation σ_i is known, it is possible to also choose the inclusion probabilities to minimize further. The diagonal term of (8) is dominant, i.e.

$$\sum_{i\in U}\sum_{j\in U}\sigma_i\sigma_j\rho_{ij}\frac{\pi_{ij}-\pi_i\pi_j}{\pi_i\pi_j} \approx \sum_{i\in U}\sigma_i^2\frac{1-\pi_i}{\pi_i}. \quad (9)$$

With the constraint of fixed sample size $\sum_{i \in U} \pi_i = n$

and by using a Lagrangian function, it follows that the minimum in π_i of the right hand side of (9) is

$$\pi_i = \frac{n\sigma_i}{\sum_{j \in U} \sigma_j}, \tag{10}$$

if each $n\sigma_i < \sum_{j \in U} \sigma_j$. As a result, a very efficient sampling design under this general model is to select samples that are well spread in the x-space with inclusion probabilities given by (10). The requirements needed in order for the samples to be approximately balanced on $f(x)$ are then fulfilled. The inclusion probabilities will not vary much within the polytopes since $\sigma(x)$ is supposed to be a Lipschitz continuous function of x. Hence, with this strategy, the anticipated variance of the HTestimator becomes small.

It is not possible to balance the sample directly on $f(x)$ since the function is obviously not known in advance. Probably, the best we can do in practise is to make sure the samples are well spread in x to have a balancing effect on the unknown function $f(x)$, and hence also have small π_{ij} when ρ_{ij} may be large.

If we use well spread probability samples together with the HT-estimator, the estimator will be very efficient (i.e. have a small variance) if the population is close to a realization of the model (6). Note also that the approach is purely design based, and the estimator maintains design unbiasedness and design consistency even if the model is false.

Example 2, given in the next section, supports the above statements. In particular, the example compares different sampling methods with respect to variance and spatial balance. It is clear that methods obtaining well spread samples are more balanced and hence produce smaller variance.

4. SOME METHODS FOR SELECTING WELL SPREAD SAMPLES

Besides spatial stratification, one of the first more novel designs for selecting a well spread sample is called generalized random tessellation stratified (GRTS), and was introduced in [2]. The GRTS design uses a specific random mapping to map two (or more) dimensions to one dimension. Basically the units are re-ordered to a list and units close in the list tend to also be close in the auxiliary space. Then a systematic πps sample is selected from the list, making sure the sample becomes well spread in the list and hence also in the auxiliary space. A drawback of GRTS is that a lot of information is lost in the mapping, especially if the space has many dimensions (i.e. many auxiliary variables). However, for two dimensions, the GRTS produces rather well spread samples.

Another idea is to map dimensions to one by use of space-filling curves, and one such design was presented and evaluated in [10]. However, we believe that mapping several dimensions to one is not the best way to achieve a well spread sample. Too much information is lost in such a mapping.

A more recent idea to achieve well spread samples is to first define a distance measure in the auxiliary space. To do so, let $\boldsymbol{x} \in R^q$ be all available auxiliary variables, where $\{1,\cdots,p\}$ correspond to the quantitative variables and $\{p+1,\cdots,q\}$ to the qualitative variables. To measure the distance between unit i and j in this q-dimensional space, [11] propose the following definition of distance

$$d(\boldsymbol{x}_i,\boldsymbol{x}_j)=\sqrt{\sum_{k=1}^{p}\left(x'_{ik}-x'_{jk}\right)^2+\sum_{k=p+1}^{q}1_{x_{ik}\neq x_{jk}}},$$

where x'_k is the standardized version of x_k. By standardizing, the auxiliary variables are approximately of equal importance. However, the above distance function is just an example and in a particular situation some other distance function may be more appropriate. Given the distace measure, the design should create a negative correlation of the inclusion indicators for close units, so that two close units seldom appear in the sample together. Such a design is not necessarily complicated. For instance, the local pivotal method (LPM) introduced in [5] is quite simple. The LPM is based on the pivotal method [4]. The main idea in LPM is to make similar units (i.e. nearby units) compete with each other for inclusion in the sample. The LPM successively updates the prescribed vector of inclusion probabilities $(\pi_1, \pi_2, \cdots, \pi_N)$ to become a vector with zeros and ones, where the ones indicate inclusion in the sample. In one step of LPM, two close units i and j with $0 < \pi_i < 1$ and $0 < \pi_j < 1$ are chosen to compete. The winner takes as much probability mass as possible from the other unit. Hence, the winner receives the new probability $a = \min(1, \pi_i + \pi_j)$ and the looser gets the new probability $b = \pi_i + \pi_j - a$. Thus, if $\pi_i + \pi_j \geq 1$, then a = 1 and the winning unit will definitely be in the sample. If $\pi_i + \pi_j < 1$, then the looser will definitely not be in the sample (since b = 0). The reduced probability vector (π_i, π_j) is updated as

$$(\pi'_i, \pi'_j) = \begin{cases} (a,b), \text{with probability } (a - \pi_j)/(a-b) \\ (b,a), \text{with probability } (a - \pi_i)/(a-b) \end{cases}$$

Now, replace (π_i, π_j) with (π'_i, π'_j). The final outcome is decided for at least one unit each update, and thus the procedure has at most N steps. In each update, unit i is chosen randomly (with equal probabilities among the units with $0<\pi_i<1$) and then its nearest neighbor $j \neq i$ (among the units with $0<\pi_j<1$) is chosen.

Another method, spatially correlated Poisson sampling (SCPS) was first described in [6] and it is a special case of the method introduced in [7]. The SCPS algorithm is a bit more complicated than LPM, but is based on the same idea. Weights are used to create a negative correlation between the inclusion indicators of nearby units, forcing the sample to be well spread. For more on the above discussed methods, for selecting well spread samples, we refer the reader to the previously mentioned papers.

The two designs, LPM and SCPS were used in [11] to obtain well spread probability samples. The fact that LPM and SCPS produce well spread samples has been justified by both theoretical results and simulation results in the previously mentioned papers. Variance estimators for the HT-estimator under well spread samples was suggested in the papers [11,12]. To our knowledge LPM and SCPS are the designs that in general produce the lowest mean value of the balance measure (1) in general auxiliary space with prescribed inclusion probabilities.

When it comes to efficiency of the HT-estimator for well spread samples, we can also make heuristic arguments that such samples produce a low variance of the HT-estimator. When the sample size is fixed, the variance of $\hat{X}=\sum_{i \in s} x_i/\pi_i$ can be written as

$$V\left(\hat{X}\right)=-\frac{1}{2}\sum_{(i,j)\in U}\left(\pi_{ij}-\pi_i\pi_j\right)\left(\frac{x_i}{\pi_i}-\frac{x_j}{\pi_j}\right)^2. \tag{11}$$

A property that e.g. the LPM and SCPS design have is that π_{ij} is small (minimum or close to minimum) when $d\left(x_i,x_j\right)$ is small and π_{ij} is large (close to $\pi_i\pi_j$) when

$d(x_i, x_j)$ is large. If π_{ij} is small when $d(x_i, x_j)$ is small, then $(x_i/\pi_i - x_j/\pi_j)^2$ is small when π_{ij} is smalli.e. $(x_i/\pi_i - x_j/\pi_j)^2$ is small when $-(\pi_{ij} - \pi_i\pi_j)$ is large. Also, if $(x_i/\pi_i - x_j/\pi_j)^2$ is large (i.e. $d(x_i, x_j)$

is large), then $-(\pi_{ij} - \pi_i\pi_j)$ is small since $\pi_{ij} \approx \pi_i\pi_j$. As a result the variance (11) becomes small.

For well spread samples, the balancing property can only be shown to hold exactly in very specific situations, i.e. under assumptions (A.0)-(A.2), see (2). For a categorical auxiliary variable, the sample will be balanced if the design produces stratification with fixed sample size for each category. A simple example follows.

Example 1. Let U be a population of males Um and females Uf. Let x be the only auxiliary variable and let $x_i = 0$ if male and $x_i = 1$ if female. Also, let $\pi_i^m = n_m/N_m$ and $\pi_i^f = n_f/N_f$ be the inclusion probabilities, where nm and nf are integers. In this special case, we have that e.g. the LPM and SCPS automatically produces stratification with fixed sample sizes. Hence we have

$$\sum_{i\in s}\frac{x_i}{\pi_i} = \sum_{i\in s_m}\frac{0}{\pi_i^m} + \sum_{i\in s_f}\frac{1}{\pi_i^f} = N_f = \sum_{i\in U} x_i,$$

where sm and sf are the sampled males and females respectively.

Example 2. We compare the different sampling methods LPM, SCPS, GRTS and simple random sampling (SRS) using a model satisfying (6). In particular, the population is generated from $y_i = f(x_i) + \epsilon_i$ with

$$f(x_i) = 3(x_{i1} + x_{i2}) + \sin\{6(x_{i1} + x_{i2})\}.$$

The population size is N = 200 and the x-values are generated from a uniform distribution on the unit square. Using Euclidean distance, the covariance function for ϵ is defined as $\text{cov}_M(\epsilon_i, \epsilon_j) = \sigma^2 \cdot \exp\{-d(\boldsymbol{x}_i, \boldsymbol{x}_j)/r\}$, which is a simple covariance function used for stationary fields [13]. The ϵ-values are generated in two steps. First random independent and identically distributed data $\boldsymbol{e} = (e_1, \cdots, e_N)'$ is generated from $N(0,1)$. Then, the ϵ-values are constructed as $\epsilon = \Sigma^{1/2} \cdot \boldsymbol{e}$ using the covariance matrix $\Sigma = \{\text{cov}_M(\epsilon_i, \epsilon_j)\}$. In this example $\sigma = 0.5$ and $r = 0.1$. The units are sampled with equal inclusion probabilities and $n = 50$ units are sampled.

The target parameter is $Y = \sum_{i \in U} y_i$. In our particular realization the true value is $Y \approx 607$. The result is presented in Table 1. A clear connection between well spread samples and variance can be observed. A design with a small expected value of B, see (1), gives better estimates. Concerning anticipated variance we get similar results if we average over repeated realizations from the model.

Table 1. Results for Example 2. Empirical variance $\hat{V}$ of the HT-estimator and the mean of the measure B for 1000 samples of size 50.

Design	$\hat{V}$	Mean (B)
SCPS	166	0.137
LPM	196	0.138
GRTS	228	0.176
SRS	1305	0.359

5. FINAL COMMENTS

It has been shown that in general there is a significant balancing effect for well spread samples. Usually, well spread samples are not as balanced on the auxiliary x-variables as samples selected by the cube method, but nearly so if the sample size is not too small. However, for target variables that are non-linear in x, well spread samples are likely to be more balanced on the target variables than samples selected by the cube method. In that way, well spread samples are good for more general situations. Hopefully, the fact that a significant balancing effect has been shown will increase the interest of using well spread probability samples when auxiliary x-variables are available.

There also exists a possibility to combine the cube method with a similar idea as used in the LPM, to have a local cube method. Then samples that are both well spread (spatially balanced) and balanced on the auxiliary variables can be selected. Such a method was developed in [9].

In [1,14], properties of spatial total estimators are studied under a tessellation stratified design in a continues universe. With similar assumptions on the target function, as used in this paper, they show that the convergence rate of the variance of the total estimator is $O(n^{-2})$ for such a design. Even though our setting is different and does not imply a strict stratification, this indicates that spreading the sample locations well probably gives a small variance when there are spatial trends.

In the setting of Voronoi polytopes used in this paper, we may consider the nearest neighbor estimator (NNestimator) in place of the HT-estimator. The NN-estimator of Y is, if n_i is the number of units in polytope p_i,

$$\sum_{i \in s} y_i n_i .$$

Under the assumptions (A.0) and (A.1), we have $n_i = 1/\pi_i$ and the NN-estimator is equal to the HT-estimator. This implies that the NN-estimator will be approximately design unbiased for well spread samples. Moreover, the NN-estimator can probably adjust for some minor spatial imbalance in the sample by using the realized polytope sizes n_i instead of $1/\pi_i$, which can be viewed as the estimated polytope sizes. The possible benefit of using the NN-estimator in place of the HTestimator will be investigated in a future paper.

ACKNOWLEDGEMENTS

Thanks to Lennart Bondesson and an anonymous reviewer for helpful comments **that improved this manuscript.**

REFERENCES

1. L. Barabesi and S. Franceschi, "Sampling Properties of Spatial Total Estimators under Tessellation Stratified Designs," Environmetrics, Vol. 22, No. 3, 2011, pp. 271- 278.

2. D. L. Stevens Jr. and A. R. Olsen, "Spatially Balanced Sampling of Natural Resources," Journal of the American Statistical Association, Vol. 99, No. 465, 2004, pp. 262- 278.

3. J.-C. Deville and Y. Tillé, "Efficient Balanced Sampling: the Cube Method," Biometrika, Vol. 91, No. 4, 2004, pp. 893-912. doi:10.1093/biomet/91.4.893

4. J.-C. Deville and Y. Tillé, "Unequal Probability Sampling without Replacement through a Splitting Method," Biometrika, Vol. 85, No. 1, 1998, pp. 89-101.

5. Grafström, N. L. P. Lundström and L. Schelin, "Spatially Balanced Sampling through the Pivotal Method," Biometrics, Vol. 68, No. 2, 2012, pp. 514-520.

6. Grafström, "Spatially Correlated Poisson Sampling," Journal of Statistical Planning and Inference, Vol. 142, No. 1, 2012, pp. 139-147.

7. L. Bondesson and D. Thorburn, "A List Sequential Sampling Method Suitable for Real-Time Sampling," Scandinavian Journal of Statistics, Vol. 35, No. 3, 2008, pp. 466-483.

8. D. G. Horvitz and D. J. Thompson, "A Generalization of Sampling without Replacement from a Finite Universe," Journal of the American Statistical Association, Vol. 47, No. 260, 1952, pp. 663-685.

9. A. Grafström and Y. Tillé, "Doubly Balanced Spatial Sampling with Spreading and Restitution of Auxiliary Totals," Environmetrics, in Press, 2012.

10. J. Lister and C. T. Scott, "Use of Space-Filling Curves to Select Sample Locations in Natural Resource Monitoring Studies," Environmental Monitoring and Assessment, Vol. 149, No. 1-4, 2009, pp. 71-80.

11. Grafström and L. Schelin, "How to Select Representative Samples," Scandinavian Journal of Statistics, 2013.

12. D. L. Stevens Jr. and A. R. Olsen, "Variance Estimation for Spatially Balanced Samples of Environmental Resources," Environmetrics, Vol. 14, No. 6, 2003, pp. 593- 610. doi:10.1002/env.606

13. N. A. C. Cressie, "Statistics for spatial data," Wiley, New York, 1993.

14. L. Barabesi and M. Marcheselli, "A Modified Monte Carlo Integration," International Mathematical Journal, Vol. 3, No. 5, 2003, pp. 555-565.

Index

M

N

O

P

R

S

T

V

W